CRUCIAL DECISIONS

**FROM CORONAVIRUS TO BASEBALL TO REFINERIES...
WHAT WE'VE LEARNED ABOUT DATA, EXPERTS, AND
RELIABLE RESULTS IN COMPLEX SITUATIONS.**

RYAN SITTON

J. WESTIN
BOOKS

J. Westin Books
100 Missionary Ridge
Birmingham, AL 35242
An imprint of Iron Stream Media
IronStreamMedia.com

Library of Congress Cataloging-in-Publication Data 2020918942

ISBN-13: 978-1-56309-475-0
eBook ISBN: 978-1-56309-481-1

1 2 3 4 5—24 23 22 21 20

To my wife, Jennifer...
Your diligence, your commitment,
and your attention to detail
made this book possible.
Thank you.

Contents

Introduction

Why We Crave Reliability

WHAT EVERYONE WANTS, MORE THAN ANYTHING, is reliability. Substitute whatever term you want—*stability, consistency, dependability*—in the end we all want it. It is the cornerstone of modern existence. Look at virtually any period of rapid economic growth in history and you'll often find this formula at work: a strong motivation to advance coupled with confidence in stability around us. Even the economic expansion during World War II exemplified this. Americans rallied their strength against the imminent threat of the Axis powers, but we had an advantage. Every country engaged in the war was motivated to win, but as Europe was gripped in the conflict and destruction, American soil felt relatively safe. With the combined motivation to win and stability at home, the United States went through its most aggressive industrial expansion in history. The *reliability* here at home drove unparalleled increases in productivity, and enabled America to emerge as the new global superpower.

Looking around us now, the value of reliability couldn't be more evident. When we start our cars, turn on our lights, drop our kids off at school, show up at work, receive our paycheck, or go to the grocery store, we *expect* everything in society to work. When some small thing happens that threatens to disrupt normal life, we panic. The mention of a possible toilet paper shortage caused a run on toilet paper. People went to the store, loaded up on every roll they could find, waited in line for hours, and stocked up with a year's worth of toilet paper.

Reliability makes our world go around. The more we have, the more we advance. When there is less, we slow down, or even go backward. Our dependency on reliability is shown in the effects on corporations, academic institutions, and government. When things are consistent, employees, students, and citizens feel good. Introduce a bit of uncertainty, and stock drops, students struggle, and citizens panic. This, in turn, reflects another critical aspect of our society: the gargantuan impact of decisions made in crucial moments.

The connection between key decisions and systematic reliability has brought modern society to a seminal moment in our history. Our reliability, in the developed world, has reached such epic levels that we can live very comfortably. One thing the coronavirus lockdown showed us is that the basic needs of Americans can now be met with only 25 percent of our population working. And while this is unsustainable, it is a testament to the ingenuity of people.

As we have increased productivity using machines and systems, the employment makeup of our population has gone through a major shift. We went from over 50 percent of our population working in basic goods and services in the 1950s, down to only 20 percent today. This allows the rest of our working population to go into other goods and services that enhance our lifestyle, improving everything from restaurants to clothing to vacations to cars. In other words, reliability has improved our quality of life in staggering proportions. However, the downside of that high degree of reliability is that we have now come to *expect* it. And this expectation has made it possible for a single bad decision, or a series of uncertain ones, to easily disrupt our entire way of life when unreliability causes panic.

History may judge our reactions (by individuals, societies, and government) to COVID-19 as some of the worst decisions in modern history. In a rapid response, governments shut down major portions of our economy, bankrupted thousands of companies, suspended schools, and put tens of millions of people out of work. This intensified an unprecedented level of fear, and decimated the reliability of our way of life. As people began to figure out that this was unsustainable, they pressed hard to return to work and the rest of life. Of course, the spread returned. Most economists agree that the struggle to balance a functioning society with prevention and treatment of COVID-19 will have lasting impacts, some speculating that it will take years, even decades, to recover.

But in failure comes the opportunity to learn. The human race has had several big leaps in our development, including the Stone Age, the Bronze Age, the Industrial Age, the Machine Age, the Space Age, and the Information Age/knowledge economy. Right now, we are on the precipice of what could be our next and possibly greatest advancement, as we use data in ways we never have, to make consistently better decisions and perform at levels we never thought possible.

The early signs of this started nearly twenty years ago when Billy Beane revamped the Oakland Athletics recruiting strategy based on a revolutionary use of data, making better decisions about which recruits to put on the team. Captured in the movie *Moneyball*, this idea spawned an entire shift in thinking, with other sports clubs, businesses, investors, technology gurus, gamblers, and cultural thought leaders working to advance the use of data to make game-changing leaps forward. In my own world, talking to leaders in marketing, operations, technology, human resources, and politics, we have tossed around the idea of *moneyballing* our processes. There have even been adaptations, attaching "balling" to the end of any word to indicate a substantial shift in the use of data. In his 2018 book, ASTROBALL, author Ben Reiter described how the Houston Astros took the original concept to new heights and won the world series in 2017.

While these advancements in sports, gambling, and iPhone apps are exciting to watch, they are even more impactful to the broader world because they give us a glimpse of what else is possible. With more powerful data analytics and advancements in our study of the field of data science, we can take performance in other areas to places we have previously never dreamed.

Nowhere is this potential as striking as in the world of reliability and complex systems. Improvements there would affect the most vital components of our society, from healthcare to education to energy to large business teams. How we make decisions and solve problems using data will drive the next evolution in human history. The question is, how quickly will we make the change? It took fifteen years *after* Billy Beane's magic with Oakland for the rest of the major league teams to accept that good scouts *and* good data analytics will beat good scouts alone.

In the early days of data analytics in sports, for example, the scouts and coaches with decades of experience (a.k.a. the experts) tended to refute the application of the geeks (a.k.a. the data analysts). In return, the data analysts had a temptation to view the experts as dinosaurs, not willing or able to get with the times. This polarization in industry, sports, or public policy pushed experts and geeks onto opposing sides of decisions for much of the past two decades. As a result, much energy has been spent by both the expert and the geek trying to prove themselves superior, rather than operating as two players with different skills on the same team. Hence, we turn to one *or* the other in a vacuum, make the best decisions we can, and either get it right or suffer a miscalculation.

While it may not have been the norm, there have already been shining examples of integration of experts and data analytics in decision making. When Jeff Luhnow and Sig Mejdal brought performance data analytics to the St. Louis Cardinals and then used it to win the world series with the Houston Astros, they took a different approach. "It's the scouting information *and* the performance information." In 2006, Sig "had developed the first iteration of a metric that sought to incorporate the reports of the club's scouts with his own performance-based algorithms, to integrate quantitative and qualitative evaluations. He called it *Stout*—half stats, half scouts."[1]

---

[1] Ben Reiter, ASTROBALL: THE NEW WAY TO WIN IT ALL (Three Rivers Press, 2018), 28.

It took a decade and much consternation to move professional sports to more consistent use of data. Therefore, the process of shifting reliability and risk analysis in larger, more complex organizations and industries may be even more difficult. After all, according to LinkedIn, the entire Houston Astros organization employs around seven hundred people.[2] Industries such as healthcare, oil and gas, water treatment, education, and mining encompass the world's largest organizations. This includes thousands of the most accomplished leaders, their most seasoned engineers and scientists, and tens of thousands of employees. A bigger challenge, yes, but also a bigger opportunity!

Since our decisions have become more complex, and affect more people than ever before, they are more crucial than at any time in history. Shifting our decision process toward quantitative methods will not only make the world more reliable but allow more people to focus on unlocking even more creative and forward-thinking ideas. This could make the quantification of crucial decisions the biggest advancement in the world of data science. This book lays out the bold ideas to start.

---

[2] "Houston Astros," LinkedIn.com, https://www.linkedin.com/company/houston-astros/.

Chapter 1

The Pandemic:
The Reliability of People

IN 2050, I WILL BE SEVENTY-FIVE years old. At that point, I envision there will be times and events that I will look back at with real clarity, i.e., key points in history. My parents remember where they were when JFK was shot. My father can recall the stress and decisions around Vietnam like it was yesterday. People reflect on the Cold War, the New Deal, and the Great Depression with thoughtful analysis as they are considered pivotal moments and events in our history. I remember the sadness when the Challenger space shuttle exploded. And the time my sister fell into a creek behind our house and I held on to her hand while I yelled for my dad. I remember standing in a bedroom in a townhome in Australia when I heard for the first time that America had been attacked by terrorists on September 11, 2001.

I imagine in a decade or so that our generation will reflect on the COVID-19 lockdown as one of our society's biggest mistakes. Why? Because I believe that we will look back and recognize that we caused unmitigated and unparalleled damage to our economy, our society, and our way of life through a combination of fear, poor analysis, and even worse decisions. Not because people weren't good. And not because they didn't want to do good things. After all, didn't we want to save lives? When I think of my aging parents, and the risks that coronavirus presented to our elder population, it scared me too.

It is rare that decisions we make are so critical that they can have a major impact on our lives. We make thousands of decisions a day—what to eat, where to walk, what to say, none of which have a lasting impact. Even big decisions—what car to buy, where to live, what job to take—feel major in the moment but generally don't register with too much command in our life story. Who to marry, what degree to study, whether to join the armed forces, whether to have a child—these typically represent our most lasting decisions, those we consider *crucial*. In today's world of instant everything, most things are less permanent. Therefore, crucial decisions are less and less frequent. Where do we see them? When leaders are making decisions in complex situations.

Back to coronavirus in 2020, did we save lives? This is a difficult question. After all, every one of us will eventually pass away. If by "save lives" we mean certain lives were extended, then yes, some were. However, what if another group of lives were shortened? When all is said and done, I believe the data will reveal this is exactly what happened. In other words, some lives were positively impacted while others were crushed, and as a society the net gain in terms of lifespan was negligible. At the same time, the economic, emotional, and mental toil of the times has been some of the worst in history. It seems that, at best, we spent trillions of dollars to prolong a small number of lives for a short time (on average). By contrast, we could have directed one tenth of those resources to improve and extend the lives of ten times as many people. More on this, shortly.

But how did we decide to prioritize these people with this extraordinary sum of money for a relatively small return? Simply put, our processes for making these crucial decisions were outdated. Leaders across the business and political spectrum were put in impossible positions, asked to make gut calls reacting to input from a small group of people, based on an incomplete set of data. This led to conclusions that drove uncertainty, not reliability.

What Happened?

"It seemed sensible enough. After we started hearing about coronavirus, it appeared that we might be facing huge problems—massive amount of sickness, high fatality rates, and overloading our healthcare system. We had to do something. The public demanded it."

Who said this? I have paraphrased, but virtually every elected official in the United States has echoed these words. You can look up quotes from across the country as states and cities were closing down. What prompted them to make this statement? Someone asked a question like, "Why did you make the decision to shut down *everything*?"

In retrospect, it's amazing how quickly everything happened. On December 31, 2019, the local government in Wuhan, China, confirmed publicly that a large number of pneumonia cases being experienced in the city were actually the effects of a new strain of coronavirus. Eleven days later, the first death was reported. Ten days after that, cases began to be reported across the world, with one case in Washington state, where a man had returned from Wuhan. Finally, by the end of January, the World Health Organization (WHO) declared a global health emergency. Thirty days passed from the first reports in China to the recognition of a global pandemic.

Over the next month, reports began to pour in. The first death outside of China was reported in the Philippines. Cruise ships were quarantined. The disease was titled COVID-19 by the WHO. The first major wave outside of China occurred in Italy and then Iran. Cases were reported across Europe. The first US death was reported on February 29, 2020. By the middle of March, most schools in the country were closed until further notice. And Venezuela and France were the first countries to impose nationwide lockdowns. March 19 was the first day China reported no new reported infections, then on March 26, the US became the largest coronavirus victim, with 81,321 cases. Through April, more cases were reported across the world and the United States. In response, state and federal government imposed restrictions and lockdowns, becoming more and more strict by the day. Job losses in the United States reached epic proportions.

On May 1, 2020, four months after the first public reports out of China, there were just over 1,000,000 confirmed cases in the United States, and around 58,000 deaths. At the same time, the lockdowns caused over 33 million Americans to lose their fulltime jobs and another estimated 10 million to lose part time jobs. With a family dependence ratio of 2:1 in the United States (for every working person there is approximately one other who depends on them financially), 86 million Americans lost their primary income. So, while 0.018 percent of Americans were killed by COVID-19, over 26 percent lost their livelihood.

Did It Work?

Time, evaluation, and retrospect will answer this question completely, but it appears now that the short answer is NO. The primary objective was to "flatten the curve." Did that mean fewer people would get the disease? No. Why do it? To ensure that hospital facilities were available to treat those who got it. The concern was that if the disease spread too fast, then too many people would need to go to the hospital at one time. However, as of May 1, 2020, the total number of people hospitalized for COVID-19 in the United States stood at 164,000[1] or an average monthly rate of around 70,000. This compares to a total of 924,107 total hospital beds in the country.[2] Plus, the federal government spent nearly one billion dollars setting up temporary hospitals with over 13,000 beds that, combined, treated just over a thousand people.[3]

By mid-June, governments were faced with the reality that their economies may never recover. Fear of coronavirus

---

[1] "Laboratory Confirmed COVID-19 Associated Hospitalizations," COVID-NET (CDC.gov), June 27, 2020. https://gis.cdc.gov/grasp/covidnet/COVID19_3.html.

[2] "Fast Facts on U. S. Hospitals," aha.org, American Hospital Association, https://www.aha.org/statistics/fast-facts-us-hospitals.

[3] "U.S. Field Hospitals Stand Down, Most Without Treating Any COVID-19 Patients," npr.org, National Public Radio, https://www.npr.org/2020/05/07/851712311/u-s-field-hospitals-stand-down-most-without-treating-any-covid-19-patients.

was being superseded by fear of losing one's house, inability to afford healthcare, or not being able to buy groceries. The fact was, ironclad lockdown worked to slow the spread, but was completely unsustainable. Hence, after three months, there were no options left. People had to get moving. As reasonably expected, when people began going from shelter in place to back to life, the virus began to spread again. By mid-July, headlines were once again focused on the number of new cases and overwhelming hospitals. In other words, three months of lockdown and trillions of dollars simply delayed the situation by three months.

There were initial promises/hopes of flattening the curve, of allowing or hoping for enough time for a vaccine to be created, tested, and implemented across the country. While this is possible, according to the Mayo Clinic, "Realistically, a vaccine will take twelve to eighteen months or longer to develop and test in human clinical trials," and then many more weeks to produce, distribute, and apply.[4] We are likely looking at two years before any serious national immunization is rolled out. So, unless we are willing to stay in lockdown for another eighteen months, very few lives will be extended due to vaccines.

Some still say that we saved lives, that without the lockdown we would have overwhelmed hospitals or exposed people who were unexposed during the lockdown. While these statements seem less and less valid, and more and more defensive in nature, here are the numbers: based on data from the first months of the disease, dozens of universities and other institutions are running models that provide projections on mortality rates, and they are combined and published by the CDC.[5] If we go back and extrapolate the realistic highs and lows of these models, one could estimate that as many as 250,000 people have been "saved" (did

---

[4] "COVID-19 Vaccine: Get the Facts," mayo.org, Mayo Clinic, https://www.mayoclinic.org/diseases-conditions/coronavirus/in-depth/coronavirus-vaccine/art-20484859.

[5] "Forecast of Total Deaths," cdc.gov, Center for Disease Control, https://www.cdc.gov/coronavirus/2019-ncov/covid-data/forecasting-us.html.

not die from COVID-19). Based on demographic statistics, the average age of these people would be 70.7 years old.[6]

Due to all our warnings and lockdowns and mandates, it appears that the biggest impact we have had is to delay the spread of the virus. The question is, who benefitted? Some did. My mother could have been one. High risk people such as the elderly have gotten more time before being exposed. At seventy-six years old, and in the final stages of Alzheimer's, my mom is very frail, and very high risk. I am certain that if coronavirus had gotten into her facility early on, she would likely have passed away. I believe that she, and others like her, will add another year or so to their lives because of all of this. The question is, was it worth it? What did we sacrifice so that my mother would be at lower risk?

What Were the Costs?

On March 1, 2020, US unemployment was at one of its lowest levels in history at 3.7%. At that time, roughly 5.8 million Americans were out of work, but there were about seven million job openings. In other words, there were more jobs than we had people who wanted them. Two months later, on May 1, over 33 million Americans had lost their jobs due to the lockdowns. This led to an unemployment level of 25.1 percent, higher than the worst year of the Great Depression (24.9 percent in 1933). Economists vary widely in their projections on recovery, from as soon as six months to as long as five years. But if we assume a nominal one-year period from the highest unemployment to the lowest, that is an average of sixteen million Americans out of work for that period. At an average annual income of $56,516 per year,[7] that is nearly one trillion dollars of direct financial loss to families.

---

[6] "Weekly Updates by Select Demographic and Geographic Characteristics," cdc.gov, Center for Disease Control, https://www.cdc.gov/nchs/nvss/vsrr/covid_weekly/index.htm#AgeAndSex.

[7] "Income and Poverty in the United States, 2015," The United States Census Bureau, updated September 13, 2016, Bernadette D. Proctor, Jessica L. Semega, and Melissa A. Kollar, https://www.census.gov/library/publications/2016/demo/p60-256.html

But the overall economic losses are even bigger. Those families who lost a job stopped driving, buying extra goods, and using nonessential services. When you add these ancillary impacts of one-fourth of the population losing its income, there is another one trillion dollars in economic loss.

In response to the unemployment level and economic slow-down, the US Congress authorized a $2.2 trillion relief package called the CARES act. It distributed money directly to US citizens, expanded unemployment benefits, and enabled businesses to get "forgivable" loans if they kept employees on the payroll. In addition, the Federal Reserve Bank announced aggressive measures to buy municipal, state, and corporate bonds, lower interest rates, and advance other loan programs to drive "quantitative easing," all in all amounting to another $2.3 trillion in investment.[8]

Between economic losses, congressional stimulus, and federal reserve programs, the response to the coronavirus pandemic will cost the United States between $4.5 and $7 trillion. For the rest of this discussion, we will use the relatively conservative figure of $5 trillion, or one fourth of our entire annual gross domestic product.

Was It Worth It?

This is a subjective question, and it will be debated for years, but if we weigh the amounts spent and the losses to some families against the projected gains by others, it seems hard to say YES, it was worth it. More than likely we have extended the life of very few people, and only by a year or two on average. Even if the number is as high as 250,000 people, at a cost of $5 trillion, this would be $20 million per person. This now begs the painful question, are those lives worth that much money? What is the value of a human life?

---

[8] "Fed Expands Corporate-Debt Backstops, Unveils New Programs to Aid States, Cities, and Small Businesses," The Wall Street Journal, updated 9 April 2020, https://www.wsj.com/articles/fed-announces-new-facilities-to-support-2-3-trillion-in-lending-11586435450.

It is nearly impossible for anyone to put a value on the life of a loved one. After all, wouldn't we spend every last dime we have to save a member of our family? However, when you are in a situation where saving some lives will damage or cost the lives of others, we are forced to weigh the options. On a large scale, society is sometimes forced to look at ways to weigh the value of life, and in some areas we actually place a value on human life nearly every day. We accept the risk of dying in a car crash so that we can earn a paycheck at work, we add safety features to cars while balancing affordability and lives saved, we set insurance rates for businesses and cars based on the possibility of injury and death, and we design roads and set speed limits knowing that lives will statistically be affected. We make decisions and set priorities every day based on some idea of the value of life. Yes, it is high, and almost everyone values the lives of children above adults, meaning your life is worth inherently more the more of it you have to live. But the value of life is not infinite or priceless, because our resources (money, time, materials) are not endless or priceless.

In fact, one of the reasons to consider the "value" of life is so that we can prioritize where to invest resources. Consider medical research. Since we do not have unlimited resources for research, as our society makes decisions about where to invest, we must consider the cost of research versus the potential number of lives that could be positively affected. That way, we ensure that our resources are going into the most effective areas.

So, what value should we use? Does anyone dare put down a number? Between looking at human behavior, insurance rates, and government reports, The Globalist[9] published a summary that said we value life anywhere between $1.5 million (the value derived in 1987 when states made the decision to increase speed limits, comparing the economic advantage of increasing speed limits compared to the increased fatality rate) and $9.7 million according to the Environmental Protection Agency. However, to

---

[9] "The Cost of a Human Life, Statistically Speaking," The Globalist, updated July 21, 2012, https://www.theglobalist.com/the-cost-of-a-human-life-statistically-speaking.

get more granular, if you divide these numbers over estimated years of life (average 78.5 year life expectancy),[10] you get a range of $19,000 to $124,000 per year of quality life.

Now down to the cold hard analysis. For my mother, this may have added a few months or a year to her life. Maybe. Let's assume that five hundred thousand people were positively impacted, double most estimates. We will assume their average age matches the national average coronavirus mortality age in the US, 70.7 years old. Our efforts extended their lives between one year (until they eventually get the disease) and eight years (the average US lifespan of 78.5 years). This means five hundred thousand to four million years of quality life. If we assume $124,000 per year of quality life, then we should have been willing to spend between $62 and $496 billion on this effort.

We spent $5 trillion.

Some will say, "Better safe than sorry," or "We did the best we could." Or my favorite, "How could we have known?!?"

To understand, we must look at an entirely different facet of modern society.

---

[10] "U. S. Data," The World Bank, https://data.worldbank.org/country/united-states?view=chart.

Chapter 2

The Plant:
The Reliability of Machines

HOW DOES AN OIL REFINERY RELATE TO CORONAVIRUS? Good question. And the answer is bigger than simply refineries and COVID-19. The underlying theories around reliability and decision-making underscore an entire host of industries and societal challenges. Another good question: What do oil refineries, water treatment facilities, mining facilities, healthcare, human resources, our education system, and COVID-19 management all have in common?

The answer is that all of these are collections of assets (yes, I am referring to people as an asset). Look at half the corporate websites in the world, which all feature the aphorism, "people are our greatest asset." Of course, consider some of the decisions made at Enron, Lehmann Brothers, and AIG, and some may say that people are our greatest liability. Regardless, the notion is true. As we look at ten thousand kids in a school district, ten thousand sick people utilizing local healthcare, a company with ten thousand employees, or a chemical plant with ten thousand pumps, pressure vessels, and sections of pipe, all are complex systems of multiple independent assets. In each case, we can easily (too easily in some cases) form opinions about one asset, but it is exceedingly hard to evaluate the system as a whole. Refineries turn out to be a good example to evaluate, for a number of reasons.

1. Refineries have been around a long time, doing the same thing—turning crude oil into useable fluids like gasoline, diesel, and jet fuel.

2. There are around 130 refineries in the United States, and around 600 in the world, and they all have the same basic function.

3. Since energy is so vital to our way of life, information on refineries is tracked and maintained in public domains.

4. While there have been a number of studies and data around education and healthcare, it is easier for most people to think objectively when talking about a bunch of inanimate machines in a plant versus kids or sick people.

Let us consider the evolution of reliability, starting with the birth of America's oil industry.

Refineries

In the 1850s, the first oil well was drilled in the United States, in Pennsylvania, and the first refinery was built in Pittsburgh. Throughout the 1800s and most of the 1900s the oil industry grew by leaps and bounds in the United States. Prior to mass production of the automobile, gasoline was not the primary product. In fact, John D. Rockefeller once referred to gasoline as an "awful waste product." Instead, for the first seventy years of the US oil industry, the primary need was kerosene. As cities were getting larger and our population becoming more urbanized, the light produced from kerosene lamps was a staple for functioning civilization.

Demand for kerosene and then later gasoline, diesel, and jet fuel all propelled the US oil and gas industry through massive expansion for nearly 130 years. By 1980, the United States was producing nearly nine million barrels per day of crude oil, and its 301 refineries were processing up to seventeen million barrels per day of crude oil[1] (including eight million barrels per day of imported crude).

---

[1] "Number and Capacity of Petroleum Refineries," U.S. Energy Information Administration, updated June 22, 2020, https://www.eia.gov/dnav/pet/pet_pnp_cap1_dcu_nus_a.htm.

However, between 1960 and 1975, thirteen countries, led by Iran, Iraq, Kuwait, Saudi Arabia, and Venezuela, formed the Organization of the Petroleum Exporting Countries (OPEC). They controlled over 30 percent of world oil production and began to tighten control of markets. In 1973 and 1974, OPEC member companies refused to send oil to the United States in retaliation for its support of Israel. This move, referred to as the Middle Eastern Oil Embargo, caused oil prices to skyrocket, more than doubling the historical maximum in just nine months. Then, in 1979, the Iranian Revolution and following Iran-Iraq War destabilized oil production, and prices shot up again. Adjusted for inflation, oil prices reached their highest point in history until being barely eclipsed in 2009. From 1973 to 1980, the high prices inspired US oil production and refineries to expand significantly.

However, after 1980, global oil production stabilized. In addition, high oil prices prompted Congress to pass fuel efficiency standards (CAFE standards), so new cars would use less gasoline. In addition, the high oil prices sparked a long period of recession and anemic economic growth, prolonged by The Cold War with the Soviet Union. This all took its toll on energy markets. Prices and margins were low and thin. The result was that the US oil industry began a long period of decline that would last nearly thirty years. Oil wells were plugged, drilling rigs decommissioned, and refineries shut down. By 2008, the US had lost over half of its refineries, and nearly half of its oil production.

After 1980, the focus of oil companies changed. While the previous hundred years were characterized by races in construction and expansion, the next thirty years featured races for efficiency. Cost reductions, operating excellence, economy of scale, and reliability became the hallmarks of a good refinery. Since you didn't have the money to expand or buy new, you had to make what you had as efficient as possible. So began the slow turn from "Fix it when it breaks," to "Fix it before it breaks," to "Keep it from breaking altogether."

Utilization

Refinery *utilization* is the rate at which refineries process crude relative to their maximum. In 1980, this rate stood at approximately 70 percent. Figure 1 shows the stated total capacity of all American refineries, the actual throughput, and the percent utilization from 1985 to 2020. As oil companies began to improve their reliability and operations over the next forty years, there would be three phases of evolution in utilization.

Phase I – 1980 to 1998—Improvement through Contraction

Phase II – 1999 through 2010—Decrease during Expansion

Phase III – 2011 through 2019—Improvement and Expansion

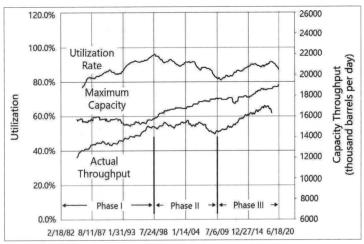

Figure 1 – Refining Utilization and Capacity

All Data sourced from the US Energy Information Agency.[2]

[2] "Refinery Utilization and Capacity," U.S Energy Information Administration, updated June 30, 2020, https://www.eia.gov/dnav/pet/pet_pnp_unc_dcu_nus_m.htm.

In Phase I, refinery utilization rose to 95 percent, largely due to shutting down smaller, less efficient plants, and streamlining larger ones. In fact, while utilization rate increased, the maximum capacity of US refineries during that time actually dropped from around 17 million barrels per day, to 15.7 million barrels per day. In other words, utilization increased largely due to consolidation, as lower performers shut down and gave way for the larger, more efficient performers to pick up the slack.

Since the start of Phase II (1999), refining capacity in America has grown steadily, to nearly 19 million barrels per day. During the first half of this run, utilization trended downward, hitting a low point in 2010 of 81 percent. While refiners continued to expand, they also retooled plants to enable them to optimize throughput to match market demand and utilize opportunity crudes—lower cost feedstock—as crude prices had begun to slowly increase. In other words, while they had larger capacities, they were working hard to make the best blend for profit, not necessarily process the most crude.

In Phase III (2011), three things happened. Domestic natural gas prices dropped well below international rates, international material and labor rates became more expensive, and US crude oil production began to increase. All three of these allowed US refineries to see big opportunities in international markets. As they have found markets for more of their products, they have maximized throughput. Utilization trended upward from 2010, reaching a high point of 92 percent in 2019.

The primary drivers behind refinery utilization have been market conditions, feedstocks, and expansion or optimization. As a result, it is difficult to interpret the effect of mechanical reliability on performance. However, there is another way to look at it.

What makes larger facilities more capable of running closer to maximum capacity? Certainly, there are elements of optimization. The more complex a facility, the easier it is to adjust production levels to match market demand and maximize throughput. However, there is something else. From 1980 to 1998, while maximum

capacity dropped by 1.3 million barrels per day, actual throughput went from 11 million to 15 million barrels per day. That is
how utilization went from 70 percent to 95 percent. How was this
accomplished? There were expansions and optimizations, making
a plant better or running it better, but there were also major reliability improvements. As people decided to spend money, they
spent it in areas that would keep them from shutting down.

Improvement

During Phase I, most of the improvements were made by evaluating what was occurring, then making better decisions as to
how maintenance and improvements should be executed. In
general, this advancement was focused largely on the acute decisions made daily in the plant, and therefore, the focus of these
improvements was on changing the way people did their jobs.

A simple example would be a pipe that thins and then leaks.
Prior to 1980, the response would typically look something like
this:

1. The section of the plant in which that pipe was functioning would have been shut down.

2. A replacement piece of pipe would be sourced, and
   unless the segment had failed recently and required a
   design change, this new section would be whatever was
   handy or a similar design.

3. The segment would be replaced, and the unit put back
   online.

After 1980, plant management began to push their maintenance staff and engineers to change the process in step two.
Instead of simply replacing the segment, they began to ask
the question, "What could be done to keep this from happening again in the future?" This could be a change in material, a
change in thickness, or even, for the more integrated companies,
a change in operations. Since failures were more prevalent in the
mid-1900s and people spent many years working in the same
facilities, this approach worked well. People were challenged

to change their approach, and with extensive knowledge of the plant and the history of its assets, there were many opportunities to make big improvements.

The common theme here was "best practice." The term began to be used in the early 1900s as the Industrial Revolution gave way to the next evolution of advancement in manufacturing, leadership principles, and modern-day business philosophies like those of Peter Drucker, "the founder of modern management."[3] After industrialists like Vanderbilt, Rockefeller, Carnegie, and Ford had built the foundations of modern society, we now needed our workers to advance their thinking and help propel industries like transportation, oil, steel, and automobiles into a new age of efficiency.

These principles all hinged on the human being as the center of the advancement. All that was needed to improve our effectiveness was for people to improve their thinking—and thereby their use of tools, systems, and decisions—and performance improvements would come. Ideas on reliability and maintenance are captured clearly in two relevant books on the subject, UPTIME and MAINTENANCE AND RELIABILITY BEST PRACTICES.

I was given my first copy of UPTIME when I was a reliability engineer at Marathon Petroleum in 2000. Its first edition was published in 1995, second in 2006. My wife received her copy of MAINTENANCE AND RELIABILITY BEST PRACTICES when she worked at Lyondell around the same time. Its first edition was released in 2008, with a second in 2013. I confess, I did not read these books at the time like I should have, but rather simply skimmed the pages so I could say that I had. They are both great books, as the authors walk the reader through the basic strategies of leading and managing a workforce in the age of reliability. One can imagine a refinery manager in 1980 buying a copy of either book for every person on his maintenance staff, as it would have represented cutting edge thinking at that time. While the books

---

[3] "The Best of Peter Drucker," Forbes, Forbes Media LLC, updated July 29, 2014, https://www.forbes.com/sites/stevedenning/2014/07/29/the-best-of-peter-drucker.

may read as somewhat dated today, they are still great guides to the basic philosophies with which people can and should approach maintenance improvement.

As evidence of the people-centric view of the time, the authors of UPTIME write, in the introduction, "How well a company executed management maintenance by using all or some of the core elements presented in this work depends primarily on how well it motivates its people." Further in that section, they write, "It is hoped that the book will be of great interest to the general manager who seeks to understand more about maintenance and for the maintenance engineering professional who wants to appreciate the bigger picture in which maintenance plays an important role. Both will gain valuable insights into the various successful maintenance management techniques and methods available today."[4]

Later, in 2008, the first edition of Ramesh Gulati's book, MAINTENANCE AND RELIABILITY BEST PRACTICES was published. Like UPTIME, the primary focus is people. In fact, in the preface to the first edition, Gulati writes, "Truly implementing a best practice requires learning, re-learning, benchmarking, and realizing better ways of ensuring high reliability and availability of equipment and systems. This book is designed to support that learning process of implementing best practices in maintenance and reliability."[5]

In short, through most of Phase I, II, and III, improvements in maintenance and reliability were viewed as the results of improvement in people. When people do improve, for the most part, that philosophy has worked well. However, the great pitfall is that the sole dependence on people will inevitably result in failure when the nature of the problem outstrips the ability of people. As we come into the next phase of reliability advancement, this is the significant challenge that is faced by large facilities and companies.

[4] J. V. Reyes-Picknell, UPTIME – STRATEGIES FOR EXCELLENCE IN MAINTENANCE MANAGEMENT, Second Edition (New York: Productivity Press, 2015), xxiii.
[5] R. Gulati, MAINTENANCE AND RELIABILITY BEST PRACTICES, SECOND EDITION (New York: Industrial Press), 2013.

The Next Phase of Reliability

As a refinery turns crude oil into a range of products, it makes money by selling those products for more money than it paid to buy and process the crude. This differential is referred to as refining "margin." While individual refineries may have drastically different margins based on regional supply and demand, access to opportunity crudes, and ability to specialize run rates to fit market conditions, there are general refining margins that provide insight into the overall refining segment.

In the three-year period including 2016, 2017, and 2018, US refining margins were, generally speaking, strong. While they did not reach peak historic levels, they remained in positive territory for most of that period, allowing refineries to turn a good profit. These good margins drove US refineries to hit a high point of 92 percent utilization at the beginning of 2019. However, this was still more than 3 percent lower than the industry's 1998 peak of 95 percent. Some of this is driven by the fact that operators and the public know more about the facilities, so "maximum capacity" is considered higher today than it was twenty years ago. In other words, the bar is simply higher, so utilization is lower despite improved equipment reliability. Regardless, when you look under the hood, there are signs that our next phase will require more than in the past. For that, we must discuss *availability*.

Availability is a measure of whether the plant process worked *when you wanted it to*. So, if you had to shut down for a planned turnaround or equipment failure, the plant or unit was unavailable. However, if you *could* run it, but decided not to because you could be more profitable with it down, then it would still be considered available. We don't often talk about industry average availability and trends, because this information is considered proprietary by most operators, and therefore is not publicly available. Regardless, it is the primary measure of effective reliability.

Since 2001, I have worked with a number of operating companies, and founded a reliability engineering and systems company in 2006. In that time, I have had the opportunity to participate in discussions and study our own data. Unfortunately, we have some limits on our data, and it is also not available for public consumption. So, after analysis, projections, and assumptions, here are our estimates, for availability ranges over the past forty years:

Table 1 – Refinery Utilization and Availability

	US Utilization Prior 10 Year Period	Average Global Refinery Availability*	Average US Refinery Availability*	Top 10% US Refinery Availability*
1980	67%	62%	74%	85%
1990	77.7%	72%	82%	89%
2000	90.3%	78%	89%	92%
2010	88.4%	81%	91%	94%
2020	87.4%	83%	92%	96%

Availability numbers after 2000 are +/- 1%,
2000 and prior are +/- 3%

There could be much debate about the nature of these availability numbers. For example, some refiners consider units that are operating at all, even at a reduced rate, to be fully available. Others do not. Also, the degree to which refineries can run past their stated maximum varies from plant to plant. This explains how availability could be lower than utilization in some instances. The most comprehensive data is gathered annually by Solomon and Associates, but once again, that information is not available to the public.

The important takeaway from this data is not the debate about accuracy, but the trends. In the twenty years from 1980 to 2000, US refineries gained 23 percent in utilization, and 15 percent in availability. Since 2000, refineries in the United States have leveled off in utilization and gained an estimated 2 percent in availability. In short, the same techniques that have gotten us this far will not get us much further.

Why is this? The short version is the 80/20 rule. 80 percent of refinery downtime was caused by the biggest 20 percent of the issues. In the early days, those issues were specific to one asset.

So, an experienced team combined with a sharp group of reliability professionals could troubleshoot the long-term problem with that one asset, make a design modification or maintenance change, and substantially improve system performance. But most of those problems and opportunities are now behind us. The next generation of improvements will have three distinct characteristics:

1. The problems and challenges are going to be much more complex in nature, spread over multiple different assets in a large system.

2. Since the performance problems are no longer being driven by one asset, but rather dozens, even hundreds of system nuances, the solutions may require multiple changes across many different assets.

3. Since these nuances will be spread over large systems and many assets, and since the problems occur much less frequently and are less obvious to human observation, they will not easily present in one set of data or measurements. We must move away from solving the problems or making the improvements we can see, to making strategic decisions based on what we can model.

How will we identify and solve these challenges? What will the next phase of advancement look like? How will it be different from what we have done in the past? The answer is the connection between coronavirus response, education, healthcare, and big industry. Gulati, Campbell, and Reyes-Picknell predicted it.

In UPTIME, they write in the Acknowledgements for the second edition, "Developments in computing technology and the explosive growth of the internet have changed the landscape for Management Information Systems dramatically, and that part of the book was clearly out of date." Later, in the section on Management and Support Systems for Maintenance, they add, "There is an array of software that manages data storage, analyzes the data, and performs reliability analyses that facilitate decisions about equipment or component replacement and inspections.

Reliability analysis makes extensive use of statistical modeling techniques, many of which are impractical to perform without a computer."[6]

In MAINTENANCE AND RELIABILITY BEST PRACTICES, Gulati writes, "Users are slowly learning the incredible power of the CMMS (Computerized Maintenance Management System) to take raw data and turn it into information and knowledge that can improve maintenance effectiveness dramatically through use of analysis tools."[7]

Despite their prescient predictions that computers and data would revolutionize our decision process in reliability, there is something they did not tell us. Only a few years ago, industries and societies fundamentally saw data analysis as a tool to be used by people in making decisions. We now see the massive opportunity, not simply in using data, but in a new philosophy. We will move away from people at the sole center of decisions with data available as a tool. Instead, we will put people *and* data both at the center of our decisions and integrate them in new and more powerful ways. This was the foundational logic behind the Oakland A's Moneyball, and the Houston Astros Astroball. This is what we will call "reliaball" for the remainder of the book.

Reliable – adj. Performing as expected or desired for an expected period of time.

Reliaball – v. Statistically optimize the reliability of a system to maximize its value and effectiveness.

But how big is this opportunity?

---

[6] Reyes-Picknell, UPTIME, 183.
[7] Gulati, MAINTENANCE AND RELIABILITY, 68.

Chapter 3

The Opportunity:
The Value of Reliaball Decisions

IF YOU WERE GIVEN FIVE TRILLION DOLLARS TO SPEND, and you wanted to have the most positive impact on human lives, how would you go about doing it? Certainly not an easy question. Some would think of free healthcare, some would point to feeding the poor, others would look to sponsor massive research programs or build renewable energy plants. If you take a moment, you will probably start asking yourself some questions. *How can I save the most lives with this five trillion dollars? Do I actually have to keep people from dying, or if I dramatically improve their quality of life, is that enough?* When you really start to ponder the possibilities, it can be overwhelming. Imagine what we could do.

It is hard to believe that America burned through five trillion dollars' worth of economic loss, debt, government support, and stimulus during three months of coronavirus lockdown. In fact, it is saddening to know that we incurred so much devastation as a people, for what is likely minimal return. The fact is, for that much money, we could have offered everyone in America free healthcare for two years. Or put another way, if you added it to our current healthcare spend, everyone in the country could have been covered for eight years. We could have made some big strides in fighting obesity and maybe even ending Type 2 diabetes. We could have invested in research to cure cancer, improved access to exercise routines, and, my personal favorite, gone on a warpath to put a big dent in the epidemic rates of anxiety and depression in the US now estimated to be as high as 20 percent of our population.[1] Or we could have simply given every person in the country $15,000.

Keep in mind, we are not talking about imaginary money. We are talking about the money we spent to hunker down and flatten the curve of coronavirus. This investment may have had an impact as small as extending the life of a few thousand people by an average of a few months or years. Since my mom is one of those who is at risk, and therefore could be one of those served by this, it is personal to me and my family. However, if she had her faculties today, I am confident she would say that any of the ideas above are better uses of those resources than extending her life in an Alzheimer's facility by a year or two.

So, what is the opportunity? Sure, coronavirus gives us a very large, recent example of a blunder in human decision making. But thousands of people are making decisions every day, based on traditional practices, expert advice, and gut feel. In an economy of $21.4 trillion (gross domestic product), massive amounts of value are being directed with these decisions, and much of it wasted. Even if we only use the government-controlled portion of that money, 37 percent, or approximately 8 trillion dollars,[2] the United States has virtually unlimited resources to make a positive impact on the world. Most of us recognize that a good portion of this money is not being directed in the most effective ways.

Imagine if we could optimize our investments to minimize risk while maximizing improvement. This kind of analysis is already being done in some areas. We see it when Amazon runs algorithms to show us the products we are more likely to be interested in. We see it in investment portfolios that are based on substantial amounts of data to balance risk tolerance with projected growth rates. The fact is, we see it in acute applications every day, but it is the more rare, complex systems, that we have yet to figure out. These are also the areas where we can create the most value if

---

[1] "About ADAA Facts & Statistics," Anxiety and Depression Association of America, https://adaa.org/about-adaa/press-room/facts-statistics.

[2] "General Government Spending," OECD, https://data.oecd.org/gga/general-government-spending.htm.

we improve systems for better performance, since these complex systems often impact the most foundational parts of our society.

These opportunities come in all sorts of different places.

Opportunities in Industry

Major industries can include everything from copper mining operations, to wastewater treatment facilities, to food and beverage production, to oil refining. In each case, many assets are linked together to form a complex system, and that system, when optimized, can be very valuable and profitable. How valuable? Since we have already explored the history of the refining industry, let's use the largest refiner in North America, Marathon Petroleum, for this example.

In the first quarter of 2020, Marathon released its financials for 2019.[3] At the end of 2019, the company had sixteen refineries in North America, with a stated capacity total of just over three million barrels per day. At that time, their stock was at nearly $62 per share, with a market capitalization of over $41 billion. In 2019, the company processed right at 2.9 million barrels per day, for an estimated utilization of 95 percent. On a refining margin of $9.94 per barrel, the company cleared over $11 billion in total margin.

When it comes to reliability, the costs most directly related to reliability are maintenance and turnarounds. Maintenance costs were not broken out in the financials, but rather included in the total operating cost number of $6.4 billion. However, they did report $738 million in turnaround costs. Generally speaking, average annual maintenance costs are roughly equivalent to average turnaround costs. Therefore, for the sake of this example, we estimate Marathon's total maintenance and turnaround costs were $1.5 billion in 2019.

---

[3] "Annual Report, 2019," Marathon Petroleum Corporation, https://www.marathonpetroleum.com/content/documents/Investors/Annual_Report/2019_MPC_Annual_Report_and_10K.pdf.

Assume Marathon were to optimize its reliability spend, including inspection, predictive and preventive tasks, removal of wasted efforts, and reduction of downtime through better performance. Suppose that the results were a 10 percent improvement (reduction) in costs, and a 1 percent improvement in utilization. Using its 2019 operating margin, this would result in an increase in corporate earnings of over $250 million (nearly 6 percent), and a $2.4 billion increase in market cap. A major financial move. And this wouldn't be the first time that Marathon saw the results of reliability. Several years earlier they experienced it during an acquisition.

BP spent years trying to sell its massive Texas City Refinery. Finally, in 2012, it was announced that Marathon would purchase the refinery, now called Marathon Galveston Bay. After removing the value of oil and product inventories, Marathon paid BP $598 million for the 451,000 barrel-per-day plant.[4] That same year, the Motiva refinery in Port Arthur, Texas, owned by Saudi Aramco and Shell Oil, finished a massive expansion. The project increased its throughput from 285,000 bpd to 600,000 bpd, making it the nation's largest refinery. The price tag for this expansion was approximately $10 billion.

Marathon added 451,000 bpd to its company's throughput for $598 million, while Motiva added 315,000 barrels for $10 billion. For its new production, Marathon spent $1,300 per additional daily barrel, while Shell and Saudi Aramco spent $31,740 per new daily barrel. Why in the world would someone spend twenty-four times as much for additional refining capacity? The answer: reliability.

In 2005, the BP Texas City plant suffered an explosion when a drum in its isomerization unit overflowed with gasoline, forming a massive vapor cloud. Fifteen people died in the explosion, and

---

[4] "Marathon to buy BP Texas City refinery for up to $2.5 billion," Reuters, Thomson Reuters, updated October 8, 2012, https://www.reuters.com/article/us-marathon-bp/marathon-to-buy-bp-texas-city-refinery-for-up-to-2-5-billion-idUS-BRE8970KG20121008.

another 180 were injured,[5] making this one of the worst industrial accidents in modern times. After that explosion, BP spent billions of dollars upgrading the facility, responding to regulators, and settling lawsuits. Over the next seven years, there continued to be small incidents and outages, all demonstrating a lack of reliability. After making major investments and improvements, BP wanted to sell the facility, but buyers were hesitant. The risks associated with the unreliability of the plant were too high.

If BP had sold the plant at even half the per-barrel cost of the Motiva expansion, it would have received $7 billion. The cost of being unreliable was at least several billion dollars, at just one facility. Imagine the value in improvements across a company, or an entire industry.

Opportunities in Healthcare

Does the following sound familiar? You aren't feeling well. You deal with it for a couple of days, but things don't get better. Finally, you make an appointment and go see the doctor. The doctor checks your weight, heartrate, blood pressure, and temperature, and listens to you breathe. He scribbles a few notes as you describe your symptoms. After that, the doctor prescribes medication, usually an anti-inflammatory, steroid, antiviral, or antibiotic, and you are on your way.

The CDC reported that people went to the doctor 883.7 million times in 2016.[6] Of those visits, 64 percent went for a chronic condition or new problem (as opposed to follow-ups or preventive care). Of those, 48 percent received an examination or screening and 78 percent received some sort of prescribed pharmaceutical.[7] Injuries, surgeries, and lab tests are

---

[5] "BP America Refinery Explosion," CSB, U.S. Chemical Safety Board, https://www.csb.gov/bp-america-refinery-explosion.

[6] "Characteristics of Office-based Physician Visits," Centers for Disease Control and Prevention, U.S. Department of Health & Human Services, updated January 2019, https://www.cdc.gov/nchs/products/databriefs/db331.htm.

[7] "Therapeutic Drug Use," Centers for Disease Control and Prevention, U.S. Department of Health & Human Services, updated January 19, 2017, https://www.cdc.gov/nchs/fastats/drug-use-therapeutic.htm.

separate from these numbers. Assuming that a majority of the prescribed pharmaceuticals went to new screenings for new or chronic conditions, this means that somewhere around half of those visits, or over 400 million, participated in the scenario described above. With almost 2.5 million people per day going to the doctor, and nearly 2 million of those people being prescribed a drug, it is shocking how little data is gathered and used to improve health.

Imagine this scenario. You aren't feeling well. After a few days, you virtually connect to a medical data science portal. Once there, you input your symptoms and vitals (which can easily be checked at home). In addition, the system asks you for your sleep conditions and quality, energy levels, and even exposure to surrounding areas and people. As you put in more data, the system drills down. After about ten minutes, and without leaving your home, the system has compared your conditions to those of 100,000 doctor visits in your immediate area in the past two weeks. It also evaluates your personal history, regional temperature changes, and reported allergens in the air. Finally, it says that you have an 88 percent probability for a reaction to elevated ragweed, and based on the statistical models, you should take a natural herb that has a 74 percent chance of reducing the symptoms. The system also says there is a 41 percent chance that you have a sensitivity to lactose that could be elevating your reaction to the ragweed, so you should avoid milk and cheese for a while. It concludes by telling you that if things don't improve in three days, to call back and speak to a doctor.

Does this sound bold and unimaginable? I hope not. The possibilities and systems exist today. We simply have to start applying them. This is for non-life-impairing conditions. Let's take something that has a more dramatic impact on life, and for which treatment is currently more extensive: diabetes.

In 2018, 34.2 million Americans had diabetes. In 2017, diabetes was listed as a cause for 270,702 deaths, with 83,564 of those listing diabetes as the primary underlying cause. In that same year, diabetes cost our society $327 billion, between direct medical costs and lost productivity.[8] Of the 34.2 million people with diabetes, 1.6 million of them suffer from Type 1, and 32.6 million (or 95 percent) suffer from Type 2. This is key, because Type 1 is an autoimmune disease caused primarily by genetic predisposition but Type 2 develops over time. According to the Mayo Clinic website, "Type 2 diabetes develops when the body becomes resistant to insulin or when the pancreas is unable to produce enough insulin. Exactly why this happens is unknown, although genetics and environmental factors, such as being over-weight and inactive, seem to be contributing factors." [9]

The CDC reported that 89 percent of those with diabetes also exhibited obesity. While there is no "cure" for diabetes, the Mayo Clinic site also states, "losing weight, eating well, and exercising can help manage the disease." While this represents the prevail-ing theory, there is another theory that rises out of an evaluation of data. According to the CDC, less than 500,000 people and 1 percent of the population suffered from diabetes in 1960.[10] Today, that number is 11 percent. Did our society develop a predisposi-tion to diabetes in two generations? Impossible.

Instead, let's look at two other data sets: the rate of sweetener use and the rate of obesity. First, let's look at a graph of diabetes rates over time.

[8] "National Diabetes Statistics Report 2020," U.S. Department of Health & Hu-man Services, updated 2020, https://www.cdc.gov/diabetes/pdfs/data/statistics/national-diabetes-statistics-report.pdf.

[9] "Type 2 diabetes," Mayo Clinic, https://www.mayoclinic.org/diseases-condi-tions/type-2-diabetes/symptoms-causes/syc-20351193.

[10] "Long-term Trends in Diabetes," Centers for Disease Control and Prevention, U.S. Department of Health & Human Services, updated April 2017, https://www.cdc.gov/diabetes/statistics/slides/long_term_trends.pdf.

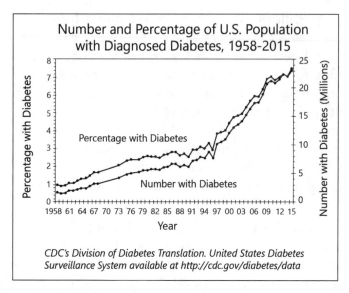

Figure 2. US Diabetes Rates

Notice the diabetes rate climbed from 2.5 percent in 1980 to nearly 7 percent in 2010. The following graph shows the rate of obesity and the rate of sugar intake per capita in the United States over a similar time.

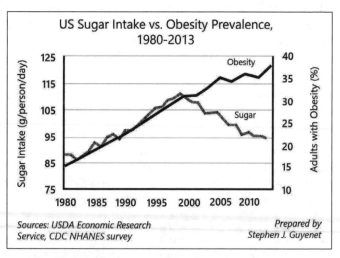

Figure 3. US Sugar Intake and Obesity Rates

In this graph, the obesity rate went from 15 percent to 35 percent, a growth rate virtually identical to the diabetes growth rate, showing a tight correlation between the two. Prior to 1980, sugar and obesity had a similar correlation. However, as you can see above, these two lines separated in the late 1990s, leading many to believe that sugar and obesity are not as closely related as previously thought. If sugar is not a direct cause of obesity, then the question remains, "Does obesity cause diabetes?"

In his book, THE OBESITY CODE, Dr. Jason Fung explains the flaw in the analysis. He spends the first part of the book explaining that despite modern beliefs, caloric intake does not have a major effect on obesity. Instead, he says, "Insulin is a storage hormone. Ample intake of food leads to insulin release. Insulin then turns on storage of sugar and fat. When there is no intake of food, insulin levels fall, and burning of sugar and fat is turned on. . . . High insulin levels cause weight gain."[11]

What causes high insulin levels? "Insulin resistance leads to high insulin levels." He goes on to explain that insulin resistance occurs when insulin receptors on cells in our body begin to "ignore" insulin, and therefore not absorb glucose. The reaction is that the body produces more insulin to overwhelm the receptors and allow cells to absorb glucose. What causes insulin resistance? The answer is high insulin levels.

A 1993 study measured this effect. Patients were started on intensive insulin treatment. In six months, they went from no insulin to 100 units a day on average. Their blood sugars were very, very well controlled. But the more insulin they took, the more insulin resistance they got – a direct causal relationship, as inseparable as a shadow is from a body. Even as their sugars got better, *their diabetes was getting worse!* These patients also gained an average of approximately 19 pounds (8.7 kilograms), *despite lowering their*

[11] D. J. Fung, THE OBESITY CODE: UNLOCKING THE SECRETS OF WEIGHT Loss (Vancouver: Greystone Books, 2016).

caloric intake by 3000 calories per day. It didn't matter. Not only does insulin cause insulin resistance, it also causes weight gain.
—Jason Fung, THE OBESITY CODE, (Vancouver: Greystone Books, 2016).

High insulin levels, therefore, cause insulin resistance, which requires higher insulin levels, whether produced by the body or injected as a medical treatment for diabetes. It is a vicious cycle. But where does it start? Sugar, especially refined sugars in raw form, can easily spike insulin levels. So, as children and young adults eat loads and loads of sugar, they are not getting fat at first, but they are slowly but surely increasing their insulin resistivity. But sugar is not directly linked to obesity, right? The trick in graph 3 above is that sugar is considered any "caloric sweetener," including honey and corn syrup. What happened around 2000? We were starting to see the effects of massive amounts of artificial sweeteners that had zero calories, like aspartame, sucralose, and stevia.

"Sucralose," Dr. Fung says, "raises insulin by 20 percent, despite the fact that it contains no calories and no sugar. This insulin-raising effect has also been shown for other artificial sweeteners, including the 'natural' sweetener stevia. Despite having a minimal effect on blood sugars, both aspartame and stevia raised insulin levels *higher even than table sugar*."[12]

So now we see the cause and the effect. No, obesity does not cause diabetes. Sweeteners, when consumed in elevated quantities over time, cause both. And herein lies the opportunity. By understanding this data, we can begin to make policy changes that drive different behaviors. For example, instead of the sugar and corn industry receiving 40 percent of our Federal Farm Subsidies,[13] perhaps we should tax those items more heavily and subsidize vegetables, proteins, and healthy grains. Whatever the

---

[12] Fung, THE OBESITY CODE, 172.
[13] "Federal farm subsidies: What the data says," USAFacts, USAFacts Institute, updated June 4, 2019, https://usafacts.org/articles/federal-farm-subsidies-what-data-says.

path, if we can reduce our rate of diabetes to 1960 levels, 30 million Americans would see a substantial improvement in their quality of life, and over 100,000 lives may be saved every year.

Whether it is a common cold, food allergies, high blood pressure, diabetes, or diet, there is so much we can learn from data if we are willing to look at the complex system of everyone out there, and start to understand the trends linking our behaviors to health. In this way, we can use data to improve the condition of human assets.

Opportunities in Education

These types of examples span all sorts of industries, government programs, and non-profits. In most parts of the country, there is a general belief that education is the key to opportunity in the world. As data poured in correlating college degrees with higher earnings, politicians went on the warpath to send every kid to college. With top-tier public schools now overwhelmed with applications, Texas implemented a "top 10 percent" rule, which meant that any child who graduated in the top 10 percent of their class in high school could go to any Texas public university. While the number of kids attending college was ramping up, the costs of education did as well. More kids were graduating with degrees and debt. Unfortunately, more kids going to college did not translate into more jobs available to pay the big salaries. Without the big salaries, kids are struggling to pay for their college debt. Now some politicians are talking about forgiving student debt.

In hindsight, flooding the market with college graduates would not fundamentally change the job market. Well, what can we do? The opportunity here is to start earlier helping kids determine where their natural skills lie. Perhaps it is studying engineering or business at a four-year university. Perhaps it is getting a medical or law degree. Perhaps it is attending a trade school to learn to be an electrician, welder, or auto mechanic. I believe there will be a new trend where employers hire eighteen-year-old kids who hate school, then train them to be strong, productive members of their team, where they can earn a substantial salary and enjoy their job. The point is, by using data around strengths and weaknesses, we can begin to help kids migrate to areas where they naturally excel and where there is future job demand. Imagine the employment landscape if more people were pursuing things that use their talents. This has a much higher probability of resulting in greater work fulfillment and better earnings, all just by doing the things we are good at.

This same logic could be applied to hiring, annual performance reviews, and promotions. How many times have you been frustrated with one of these? The fact is, despite numerous processes and ever-increasing HR systems to alleviate bias, it is impossible. By the time most of us are managing and hiring other people, we have been on this world around thirty years, and have formed a massive level of preconceived notions that operate just below our conscious decision making.

In the hiring process, someone fills out an application, then comes in for one or more interviews, in which one or more people subjectively evaluate the other's potential to do well in a job based on a conversation about prior experiences that we try to make relevant to the job at hand. When you say it out loud, there is no wonder that 38.2 million people (nearly one-fourth of the workforce) changed jobs in 2017, and 40 percent of those left within twelve months of getting the job.[14]

In this day and age, the old methods are more than old; they are archaic.

Imagine an interview process in which you showed up on-site and had already taken a battery of third-party tests that show you have the knowledge, cognitive skills, and motivating factors to do a job. Then, based on these results, the recruiter said the company wanted to talk to you about two positions that you appear to be

---

[14] "Why Are Workers Quitting Their Jobs in Record Numbers?," SHRM, Society of Human Resource Management, updated December 12, 2018, https://www. shrm.org/resourcesandtools/hr-topics/talent-acquisition/pages/workers-are-quitting-jobs-record-numbers.aspx.

a good, statistical fit for. You spend some time with people who are doing the job and learn what will be included. At the end, you meet with the hiring manager and talk about what you saw, how you would do the job, and what each of you would expect of the other. In other words, let the systems evaluate the probability that a person will be a good fit for the job, and then the candidate and the hiring manager can focus on whether they make a good connection and would like to work together.

Sound unbelievably easy? We have this ability today. We just don't use it.

Can It Work?

Yes. Despite the fact that experts across a range of fields will state the contrary, just like the experienced baseball scouts argued with the data analysts in the late 1990s and early 2000s. Back then, a generation of men had grown up looking at high school baseball players and had developed a long list of criteria with which to make personnel choices. There was just no way that a computer, some fancy models, and a bunch of data they already had could make better decisions. However, by the time Andrew Luhnow left the St. Louis Cardinals to become the Houston Astros General Manager in 2010, the debate had shifted. While Luhnow understood the risks, there was no question that using data was the way to win. Everyone was moving in the new direction. The only question was how to integrate the scouts, the experts, in the modern-day strategy as fast as possible. Then, when the 2014 "L-Astros" went on to win the World Series three years later, the proof was in the pudding.

Whether an engineer is trying to upgrade a compressor in a chemical plant, a kid is trying to select a college degree, or a patient is trying to determine the right combination of diet and lifestyle to improve health, applying a wide swath of data with better algorithms has the potential to reliaball almost any facet of our world. The only question is, how do we integrate the experts with the data analytics?

Inherent in all these shifts are "risks." One of the most common human biases is a predisposition to the current path. Another is fear of the unknown. That is, most people perceive the risks of change or the risks of the unknown as paralyzingly large. The fear of these risks can either keep people from trying anything new or prompt them to shut down an entire nation. Inherent in all of this is an outsized perception of risk. Therefore, before we can reliaball any crucial decision, we must recalibrate our ability to quantify risk.

Chapter 4

The Risk:
Probability and Consequences

Whoa, whoa, whoa . . .
you can't put a dollar figure on human life.
—Anyone

WE PUT A VALUE ON HUMAN LIFE EVERY DAY. How? When we drive to work, there is a chance, however small, that we could die in a car accident. When we decide to drive to work anyway, we are – without even realizing it – placing a value on our life. If it were true that our life was invaluable, then no paycheck would be big enough to take that chance. Instead, we decide that it *is* worth the risk, meaning that the paycheck is worth more than the increased chance at living when not driving in the car. When we do the math, we determine that the average commuter's life is worth less than $387 million. Here is how we get that number:

- 37,473 people died in car accidents in the United States in 2017.[1]

- 115,500,000 Americans commute to work every day.[2]

- Approximately 45 percent of the miles driven in the US are to work, so 16,863 deaths are attributed to work driving.

---

[1] "2018 Fatal Motor Vehicle Crashes: Overview," National Highway Traffic Safety Administration, U.S. Department of Transportation, https://crashstats.nhtsa.dot.gov/Api/Public/ViewPublication/812826.

[2] "America's commuting choices: 5 major takeaways from 2016 census data," Brookings, The Brookings Institution, updated October 3, 2017, https://www.brookings.edu/blog/the-avenue/2017/10/03/americans-commuting-choices-5-major-takeaways-from-2016-census-data.

- 16,863 / 115,000,000 = 0.0146 percent chance of dying.
- Average US per person income: $56,516 per year.[3]
- $56,516 / 0.0146 percent = $387,000,000 maximum value of life.

This, of course, is based on an average commute time and average salary levels. To someone working for $15 per hour, commuting sixty miles each way, this number drops to almost $50 million.

Explanation of Risk

Risk is a common way of evaluating the potential loss, damage, or failure of an event. By quantifying risk, people can determine if actions or ventures make sense, or if the potential gains outweigh the risks. For example, when someone invests money in the stock market, they perceive that the probability of their investments increasing in value outweighs the probability of them losing value. By contrast, if someone says, "That's a risky investment," they usually mean there is a greater probability of losing value. However, in those scenarios, the investor must also perceive that, while there is a greater probability of losing value, there is also a possibility of gaining much higher value. So, they are willing to accept the higher risk, given the greater reward. Broadly speaking, risk is calculated using the notional formula: RISK (associated with an event) = PROBABILITY (of the event) x CONSEQUENCE (of the event).

This formula is the basis of virtually every insurance product in the world. When a driver gets car insurance, the insurance company applies this formula to determine the rate the customer must pay. If a driver gets into a major accident, there will be cars damaged or totaled, injuries that need medical treatment, and potential loss of wages while someone cannot work. Most companies have average numbers for this, but let's use a total average

---

[3] "Income and Poverty in the United States: 2015," United States Census Bureau, U.S. Department of Commerce, updated September 13, 2016, https://www.census.gov/library/publications/2016/demo/p60-256.html.

value of $400,000. The probability of this event occurring can be examined by looking at the number of drivers in the US and the average number of major accidents every year. For simplicity, let's assume there are 500,000 major accidents in a year, and there are 200,000,000 drivers. Now we can calculate the risk:

$$Risk = Probability \times Consequence$$

$$Risk = \frac{500,000 \text{ accidents}}{200,000,000 \text{ drivers}} \times \$400,000 = \$1,000 \text{ per year per driver}$$

As long as the insurance company has enough drivers paying premiums, and the average premium is greater than $1,000, the insurance company should make a profit for insuring against major accidents. To develop a complete insurance policy, the insurance company must consider the risks due to minor accidents, vandalism, and theft, which are additional risk calculations. Furthermore, we don't all pay the same rates because we don't have the same risk. A college-educated mother who is 40 and drives a minivan exhibits a much lower probability of an accident than a sixteen-year-old male driving a Camaro. The insurance companies have good data on this, and apply this data to calculate actual risk, and differentiate premiums for different drivers.

In short, we use data to gain more specific insights and narrow down risk.

Calculating Risk and Reliability

If someone says you are a "reliable" driver, what does that mean? That you show up on time? That you are safe? That you are *low risk*? Probably any of these. However, to be more technical, risk and reliability are related, but not opposites on the same scale. Risk is a measure of probability and consequence, whereas reliability is a measure of cumulative probability over time, regardless of consequence. So, they are connected in probability, but risk incorporates the probability of a single event; reliability considers the amount of time that the event does not occur.

Technically speaking, reliability is calculated in time, like this:

$$\text{Reliability} = \frac{\text{amount of total time}}{\text{number of incidents that occur}}$$

$$\text{Probability (over a given period)} = \frac{\text{number of events}}{\text{time in the period}}$$

$$\text{Or Probability} = \frac{1}{\text{Reliability}}$$

For example, if the average driver gets into five car accidents (of any kind) over a twenty-year span, then:

$$\text{Reliability} = \frac{20 \text{ years}}{5 \text{ accidents}} = 4 \text{ MTTA (mean time to accident)}$$

$$\text{Probability (accidents per year)} = \frac{5 \text{ accidents}}{20 \text{ years}} = 25\% \frac{\text{accidents}}{\text{year}}$$

Using this as a basis, we could say that a driver who goes ten years without an accident is a reliable driver, and that her probability of an accident is relatively low. She should receive lower car insurance rates as a result. This concept is important because, in order to optimize a system or set of assets, we cannot stop at simply measuring and improving reliability. Otherwise, we could make the mistake of overcompensating when additional reliability is not valuable. We must consider risk, and therefore probability. If analyzed correctly, as investments are made to improve reliability over a period, risk should be reduced by a greater amount over that period. Otherwise, the improvement was a poor investment.

Here is some simple math to help explain for a plant scenario. Assume that while a plant is running, it makes $1 million per day in margin. Over the last few years, the plant has experienced an average of thirty-six days of downtime per year, driven by six failures per year. The plant runs seven days a week, so its reliability is:

$$\text{Reliability} = \frac{365 \text{ days}}{6 \text{ incidents}} = 60.8 \text{ days}$$

From the math above, there is a 600% chance of failure in one year (six per year), and the plant comes down for six days per failure at $1 million per day. Currently, the risk is:

Risk = Consequence x Probability

= (6 days x $1,000,000) x 600% = $36,000,000 in Risk

Now, if we want to improve the performance, and reduce the risk, we can improve the reliability, and lower the probability. If the plant could run for 121.6 days without failure (three failures per year), the annual risk would be only $18 million. As long as the investment is less than the reduced risk, it is a good decision. This can get complicated as we consider the long-term value of money, so for the sake of this example, if we expect a minimum of a five-year payback, our investment limit is $18 million per year x 5 years, or $90 million total.

This analysis is a basic one that is mirrored for businesses and operations around the world. From real estate to medical, from banking to construction, and from water treatment to farms, everyone is looking at how to invest resources to either expand or improve. That improvement, especially in a complex system, is often centered around reducing risk by improving reliability. But the key here is *optimization*, or maximizing the balance between resources invested and improved performance. When you have maximized the amount of gain (reduced risk) for the least amount of investment, you have optimized both reliability and risk. In other words, you reliaballed it, and this should result in the greatest overall value.

This is explained here for one key reason. Leaders around the world are struggling with this today. Many overestimate risk and invest far too heavily to manage it. Others underestimate the probability of an issue associated with certain factors, people, or assets, and don't adequately address them. Regardless of the miscalculation, the problem is almost never the judgment of the leader. It is the lack of an effective system to inform the decisions given the complex data in front of them. Most systems they do have contain two key gaps. First, their decision process is based on a single point or event in a system, not the system as a whole. While this enabled major advances through the last century, looking at the system as a whole requires a much more complex analysis. Second, much of the analyses that have been done use data but are still based on human estimation and opinion. As problems have gotten more complex, incorrect estimations of risk and reliability by people have introduced miscalculations and led to poor decisions, bad investments, and inadequate reliability.

Coronavirus and Risk

In Chapter 1, we calculated the potential values of life saved and the amount of money invested. However, the real failing of leaders and people was in understanding and calculating risk. In other words, it was hard for medical professionals or government leaders to quantify the rate that infections would spread, the percentage of people that would need to be hospitalized, or the probability that people would die. It also now appears that no one was analyzing the risk to families and individuals of a massive lockdown and economic destruction. In summary, it was decisions by people, getting information from other people, all while weighing only the desire to limit the spread, without weighing the impacts of major actions.

Here is how it happened. At the beginning of February, federal, state, and local officials began hearing about coronavirus spreading around the world. In addition, there were reports of some severe reactions. Between the infection rate and the mortality rate, people were starting to get very concerned. By the beginning of March, reports began to come out about Italy's situation—major mortality rates, hospitals struggling to keep up. Then, just a couple of weeks later, New York was experiencing rapid exposure. On March 24, Governor Cuomo announced that the city might need 140,000 hospital beds, and they only had 53,000.

By March 21, over 300,000 people had contracted the disease worldwide, and medical facilities were gathering loads of data.

What was happening with that data? Universities were putting the data into models to predict numbers of infections and deaths, ranging from 200,000 to 2,000,000. Meanwhile, medical professionals around the country were being interviewed and giving comments. From state health officials to Dr. Anthony Fauci, the thirty-six-year director of the National Institute of Allergy and Infectious Diseases, experts were all over the news, trying to balance a calm approach with the possibility of a truly awful situation.

News stories, media outlets, and social media went viral with stories of people getting the disease. "I just heard that a four-year-old without any prior conditions just died from COVID," my wife commented one evening. Families and friends around the country joined into discussions about the potential pandemic, terrified that someone may get this deadly virus, and that there would be no hospital bed or ventilators to save them. Public polls showed that 70 percent of people wanted some action by government, and right in the middle of an election year.

Imagine that you were a county judge, state governor, or the President. With medical professionals, news media, and the general public all screaming concerns, what can you do? Yes, there were early recommendations of social distancing, but numbers around the world continued to climb. Then, a clever catchphrase made its way into headlines: "flatten the curve." Everyone was saying it, while few knew exactly what it meant. Regardless of which elected official issued the order, social convention basically drove us to it. By April, the country was under lockdown.

At the time, I was serving as Texas Railroad Commissioner, a statewide elected position regulating oil, gas, and coal mining operations. As such, I had relationships with many other elected officials, and talked to several of them through the crisis. One thing struck me. No one was *really* looking at data. Sure, there were pockets of information and data out there, but no one knew which ones to trust. No one knew how to use them to make decisions. In fact, in the social and political echo chambers around this pandemic, everyone was referencing some expert, some news story, or some anecdote. In the end, as people closed down their business, stopped going to work, laid off employees, or shut down their city, county, or state, they muttered the same words.

"It is about saving lives." Simple. Noble. Completely inadequate.

As a society, we treated this like a medical problem. Virtually every news story referenced one medical doctor or another, specializing in everything from infectious diseases to emergency care. To be sure, we can assume that our finest medical experts weighed in. However, if one reads the Hippocratic Oath (or the modern-day version used by most medical schools), there is nothing about data analysis, weighing options, or valuing risk to some lives against risk to others. However, it does say, "I will prevent disease whenever I can, for prevention is preferable to cure."

We treated this as a medical problem when it was actually a data and risk problem. Yes, people were looking at the data and using it to feed the models. Where were the risk models that showed actual risk to various parts of society by age, condition, and location? Where were the risk models showing the potential economic destruction associated with lockdown? As a state regulator whose job it was to regulate the oil business, I couldn't find any model that adequately showed the impact on oil demand, so I built my own.

Meanwhile, the public and elected officials were left hearing medical professionals warn about possible doomsday scenarios, echoed by media outlets who are incentivized to provide the most sensational headlines to capture readers' attention.

It is no wonder that such a massive failure in decisions occurred.

It was a data and risk analysis problem for which there were no data and risk analysis experts being quoted, which teaches us one of the most important lessons. If you are going to get an expert, make sure you get one with the right area of expertise. Because in the end, managing risk is not about understanding the worst-case scenario. Calibrating risk means calibrating uncertainty.

Uncertainty and Risk

People often say that anticipating something bad happening is worse than the experience itself. It's even worse when you aren't sure if or when something bad will actually happen. This is the origin of the saying, "Get it over with." The idea is that waiting and wondering can often be worse than the pain of the experience. From a mathematical perspective, if we *know* that something is going to happen, and *know* when it is going to happen, then risk no longer exists. Because the probability of failure is zero until the moment it happens. Therefore, as challenging as the situation may be, we simply must plan for how to deal with the inevitability.

This is often a confusing idea, especially in the world of machinery. Let's use the example of a piece of pipe. Most pipe is at risk of failure due to thinning. As fluid moves through the pipe, small amounts of metal are taken away by the fluid through physical wear (erosion) or chemical reaction (corrosion). While thinning rates are usually very low, there can be circumstances where it can happen rapidly. To many people in industry, elevated thinning levels are synonymous with higher risk, but this is wrong. If one pipe is corroding relatively quickly, but we know exactly how fast it is thinning, we have an accurate prediction of when it must be replaced; therefore, risk is low. In another situation, a pipe may be corroding at a relatively low level most of the time, but the rate fluctuates. As a result, there is a wide range of time when the pipe may fail. It is within this range that we try to calculate a probability of failure, and therefore, risk.

Remember in Chapter 3 when we covered the dirt-cheap sale of the BP Texas City refinery? Marathon bought that facility for pennies on the dollar relative to the expansion that was completed that same year at the Motiva refinery. Why? *Risk.* The fact was, given the history of the plant, there was too much uncertainty in the plant's performance, and therefore risk was too high.

In the coronavirus lockdown, no one effectively calculated risk. They calculated overall projected infection and mortality rates, and even gave ranges from 200,000 to 2 million deaths if nothing was done. But this was never converted to risk. Could it have been done? What would that have looked like? All we knew was that some groups projected ranges of numbers (mostly cases or fatalities) and left medical experts and government officials to make decisions from there. This uncertainty required leaders to make a call, to try to minimize risk without really understanding the uncertainty in the numbers or the recommendations. But managing risk without identifying uncertainty is like taking a road trip without a map. You may generally head in the right direction, but it will take a lot longer, cost a lot more, and you may never get there. In a world of uncertainty, there are two options: calculate risk given the uncertainty that exists or gather more data to reduce uncertainty.

Back to a piece of pipe. We can measure the thickness of a piece of metal using an ultrasonic thickness (UT) meter. These devices emit an ultrasonic wave that travels back and forth and can identify the thickness based on the time it takes to travel. These devices have been in use since the 1960s, and began to be used regularly to measure pipe thickness in the 1970s. The evolution of pipe inspection using these devices is instructive.

For the first hundred years of the oil business, inspections were done on piping mostly during manufacturing, to ensure quality. If pipes broke in service, they were simply replaced. After all, if something was causing pipe to thin from the inside, there wasn't a reasonable way to look for it. That was, until the UT meter became available. Starting in the 1970s, plants began using the devices to inspect both for quality in construction and to evaluate causes when there were failures or concerns of failures. For oil companies, reducing risk has always been a high priority. The American Petroleum Institute (API), the conglomerate organization for oil companies, develops best practices to be used across the industry. In 1993, they published the first edition of API 570, *The Piping Inspection Code*. This document made it standard practice to perform external visual and UT measurement inspection at regular time intervals. The concept was that everyone should gather more data, because gathering more data would reduce uncertainty, and therefore, reduce risk.

After twenty-five years of regular thickness measurements, and seven years using API 570, a new idea was born—risk-based inspection, or RBI. The idea was that inspections, the gathering of data to reduce uncertainty, should be driven not by the number of times the earth revolves around the sun (years), but rather by the risk of failure that was identified or calculated for an asset. API 581 and 580, the recommended practices covering risk-based inspection, were published in 2000 and 2002, respectively. While some oil companies championed the new approach, others were slow to adopt. Some of the most sophisticated companies have embraced the method, but in my experience implementing RBI programs for nearly twenty years, I would say that the majority of US process facilities (including refineries, chemical plants, water treatment plants, and mining facilities) are still in the process of moving to a more analytical way of prioritizing data. Still, the monumental move to data gathering and analytics as a primary tool to gauge and reduce risk has been steady, and valuable.

If this would have happened during coronavirus, imagine how much better off we might be today. After all, as of July 2020, it appeared that we were right back where we were just four months earlier. Once again, much of our population was in fear as numbers of cases were spiking, news stories were talking about potential overrun of hospitals, and governments were imposing a new round of lockdowns. The first four months of shutdowns, massive economic destruction, and unprecedented unemployment simply lengthened the timeline.

If, in fact, a large portion of our population is going to get the disease, and it is merely a matter of time, then wouldn't it be better to quantify the risks and balance with the hospitals? For example, we know that certain demographics are very likely to experience COVID-19 without major issues. If people under the age of 55 with no cardiovascular preconditions have a one in 100 chance of hospitalization, and when they do go to the hospital they spend an average of ten days in the hospital, then in Texas (with 20,000 beds available as of June 15), we could handle 2,000 patients per day, which is 20,000 new cases per day. In other words, instead of simply trying to slow the spread, let's try to balance the risk associated with spread with the risk associated with massive economic shutdowns. While this is scary, maintaining lockdowns for months and years may only help a few for a short while, and make the entire nation suffer a massive setback. In terms of risk analysis, if we aren't actually lowering the risk, then a massive investment to improve reliability is a bad one.

Clearly, what we did in March and April of 2020 was not effective. How did we get it so wrong? It was the failure to understand risk.

The Failure in Risk

Risk cannot fail. Calculating, applying, and managing risk can. I have seen it firsthand, and not just when I watched our country shut down for a massively overestimated perception of coronavirus risk. I have seen it over and over again in my career serving large industrial facilities. Let me give you an example.

My team and I were working for a large chemical plant in the Midwest. We were implementing RBI for their stationary assets, including piping, heat exchangers, pressure vessels, and pressure relief devices. All in all, there were around two thousand individual pieces of equipment. At the outset of our project, I knew that we would have a challenge. The plant's reliability engineer felt that our analyses could not replace the knowledge the plant staff had acquired with decades of experience running the facility. In particular, he referenced a situation from a few years prior in which a software program designed to calculate corrosion rates had given a bad prediction. He acknowledged that it was based on erroneous data entry, but regardless, they did not inspect on the regular interval they had in the past. Twelve months later, the pipe failed and caused a week of downtime. "When you let the computers do the thinking for you," he said, "you are going to miss something." While he participated in good faith, he raised the concern levels of everyone involved in the project, and as we captured the input of personnel into the process, it was mostly hyper-conservative. In the end, these opinions skewed our risk analysis too high, and the analysis resulted in a much heavier inspection and maintenance program than was appropriate. This wasted money and time that could have been focused on overall plant improvements.

This anecdote is similar, in small scale, to the corona-virus reaction. In both cases, overestimation of risk led to over-conservative actions and, in the case of COVID-19, drastic economic challenges. But these cases are also in the same group as the 2008/2009 financial crisis, the oil collapse of the 1980s, and the stock market crash that caused the Great Depression. In all these cases, people were evaluating risk. While some were using quantitative models, and others used human assessments, the same problem is evident: they incorrectly assessed risk—either too high or too low—causing massive losses. These examples reveal the most common cause of risk management failure – a bad assessment. In his book, THE FAILURE OF RISK MANAGE-MENT, Douglas Hubbard explores the topic of risk analysis and its shortcomings in detail.

> *The ultimate common mode of failure would be a failure of risk management itself. A weak risk management approach is effectively the biggest risk in the organization. If the initial assessment of risk is not based on meaningful measures, the risk mitigation methods—even if they could have worked—are bound to address the wrong problems. In the worst case, the erroneous conclusions led the organization down a more dangerous path that it would probably not have otherwise taken.*
> —Douglas Hubbard, THE FAILURE OF RISK MANAGEMENT

So, what causes the failures in risk analysis and management? In a word, people. In Hubbard's book, he explains the seven major risk problems to be addressed, and they are:

1. Confusion regarding the concept of risk.

2. Completely avoidable human errors in subjective judgments of risk.

3. Entirely ineffectual but popular subjective scoring methods.

4. Misconceptions that block the use of better, existing methods.

5. Recurring errors in even the most sophisticated models.

6. Institutional factors.

7. Unproductive incentive structures.

Of all these, only number 5 is about the calculations. The rest are based, in one way or another, on how people apply risk analysis. Why are people the problem? Because we are biased. In fact, we are so biased, humans are virtually incapable of making a truly objective assessment of a situation when we do not have overwhelming data and information to inform that decision. Research into decision making was pioneered by Daniel Kahneman and Amos Tversky in the mid 1900s, and they became the godfathers of Behavioral Economics. We will explore this topic in more detail in the next chapter.

Hubbard continues, "for almost all operational and strategic risk assessments in business, someone who is deemed an expert in that area is asked to assess a probability either directly or indirectly." After exploring the fact that expertise is based on a sum of experiences that cannot be systematically recalled or quantitatively analyzed, he concludes:

> As a result, it turns out that all people, including experts and managers, are very bad at assessing the probabilities of events—a skill we should expect to be critical to proper assessments of risks. The good news is that, even though research

shows some profound systematic errors in the subjective assess-
ment of risks, relatively simple techniques have been developed
that make managers fairly reliable estimators of risks. The bad
news is that almost none of these methods are widely adopted
by risk managers in organizations.[4]

This is exactly what happened when I was in the chemical plant, and what happened to our country in 2020. Experts were misconstruing experiences, and small amounts of anecdotal information, to ratchet up concerns and inaccurately project risk. In short, their perceived level of risk was based on an uninformed perception of probability. In fact, when it comes to risk in coronavirus, I don't believe anyone ever calculated it. We got confused, hearing that people were modeling the spread of the virus. In fact, this was modeling a range of *possible* consequences, and the general public was led to believe that the probability was virtually 100 percent.

The core problem was, as Hubbard describes, a failure in risk assessment, a failure driven by experts who all had one thing in common. Experts are people.

---

[4] D. W. Hubbard, THE FAILURE OF RISK MANAGEMENT: WHY IT'S BROKEN AND HOW TO FIX IT (Hoboken: John Wiley and Sons, 2009).

Chapter 5

The Expert Problem:
Why We Need Data

I F YOU ASK ANYONE WHO THE MOST CRITICAL "EXPERT" was in the United States during the coronavirus pandemic, most people will say Dr. Anthony Fauci. Virtually unknown prior to March 2020, Fauci became a household name as he was on TV with President Trump daily for briefings on the status of the virus. Media outlets featured Fauci in stories across the country at the rate of hundreds of times per day. His voice was the preeminent reference to what was happening with the virus and its spread. I have never met Dr. Fauci, but the two people I know who do know him speak very highly of his intellect and his commitment to serving the public. I don't think we can question his resume or his intelligence. The bold question that should have been asked is, "Is he an expert?"

On the website of the National Institute for Allergy and Infectious Diseases (NIAID), of which Dr. Fauci has been the director since 1984, his bio reads:

Dr. Fauci also is the longtime chief of the Laboratory of Immunoregulation. He has made many contributions to basic and clinical research on the pathogenesis and treatment of immune-mediated and infectious diseases. He helped pioneer the field of human immunoregulation by making important basic scientific observations that underpin the current understanding of the regulation of the human immune response. In addition, Dr. Fauci is widely recognized for delineating the precise ways that immunosuppressive agents modulate the human immune response. He developed effective therapies for formerly fatal inflammatory and immune-mediated diseases such as polyarteritis nodosa, granulomatosis with polyangiitis (formerly Wegener's granulomatosis), and lymphomatoid granulomatosis. A 1985 Stanford University Arthritis Center Survey of the American Rheumatism Association membership ranked Dr. Fauci's work on the treatment of polyarteritis nodosa and granulomatosis with polyangiitis among the most important advances in patient management in rheumatology over the previous 20 years.[1]

In summary, Dr. Fauci is arguably one of the world's foremost experts on treating a large number of diseases. After hours of digging into references, research, and citations, I could not find one accolade as to his expertise in the *spread* of disease. So, if our question was, how will this virus spread, and what can and should we do to prevent it, Dr. Fauci was not an expert. In fact, in the case of this particular pandemic, I'm not sure if any of the doctors that were being quoted around the country were. As I said in Chapter 4, while treating the disease was a medical problem, analyzing the risk associated with the spread was a data analysis problem. We had the wrong experts.

This example is not unique to coronavirus. Every day we make decisions using the advice of experts. While we may not call them experts, these are the people whose opinions we perceive to be informed and/or intelligent enough to provide meaningful input on a decision. This could be due to a position of authority like a supervisor or coach, a position of trust like a parent or teacher, or a position of respect like an author or thought leader. Unfortunately, instead of taking the time to research information and make our own informed decision, many of us have fallen into the habit of accepting input, perhaps much easier than we should. This leads to *herd mentality* or *groupthink*.

---

[1] "Anthony S. Fauci, M.D.," National Institute of Allergy and Infectious Diseases, U.S. Department of Health & Human Services, updated February 5, 2020, https://www.niaid.nih.gov/about/anthony-s-fauci-md-bio.

Before casting all experts under the bus, let us be clear. Experts are valuable and important. We cannot build new models to calculate risk and reliaball anything without them. However, if we are going to get the best results, we must understand the strengths and weaknesses of our experts. To understand that, we must first explore what makes people tick, including what motivates us, and how we make decisions.

Human Motivation

In 1943, a psychologist named Abraham Maslow published a paper that identified one of the foundational principles in the evaluation of human motivation. Entitled "A Theory of Human Motivation," the paper outlined Maslow's ideas on the basic structure of motivation, including happiness, for every person. It included five levels, in order: *physiological* (food, shelter), *safety, love, self-esteem*, and *self-actualization*. Over the next decade, Maslow would expand his theory, and in his 1954 book, MOTIVATION AND PERSONALITY, he provided more detail. But the crux of his theory has been in place for nearly eighty years, and it has become known as *Maslow's hierarchy of needs*. It is represented most commonly in the form of a triangle graphic, with the basic needs at the bottom, and the more complex needs at the top.

Figure 4 – Maslow's Hierarchy of Needs

Sustenance is the basic physiological needs we all have, such as air, water, food, shelter, and clothing. Safety and security includes more than just absence of danger, as a home, health, employment, and a family all contribute to our feeling secure. Love and belonging includes friendship, intimacy, and sense of connection. Self-esteem is confidence and respect of others. Self-actualization is acceptance of one's self, a self-assuredness that comes from being moral and faithful, and a feeling that one is creating and building the world around them.

While humans are very complex beings, our *needs* can be simplified to these basic categories. As we satisfy these needs, we achieve happiness, joy, and fulfillment. One of the critical components to understanding Maslow's simple structure is the idea that it is a *hierarchy*. In other words, there is a priority and a flow. In general, one cannot work to achieve love and belonging if one does not have enough to eat or is in danger. Likewise, one cannot begin to explore self-actualization if one struggles with love or esteem. This may seem elementary, but it is a vital piece of the puzzle in understanding what makes people tick.

Imagine you are asked to participate in something as an expert. What is your motivation? Assuming you are not concerned about sustenance or safety, you would like to achieve self-actualization, but before you can achieve this, you must ensure you have belonging and esteem. How does an expert achieve belonging and esteem? They want to connect with those around them, feel that they are working on a common goal, and then provide information or advice that is perceived as valuable. Take the case of a doctor treating a patient. They want the patient to feel better, and they want to believe that the patient respects and appreciates them for their treatment. In the case of coronavirus, if I was a doctor offering my opinion on what to do, I would want to make sure I was focused on saving lives and talk about how to do that. At no point would I be motivated to provide advice other than caution, and since I do not have data analytics or economic expertise to offer, my source for esteem is my medical opinion around prevention and treatment. Hence, I am an expert in one piece of this pandemic, but not the entire issue. Given the motivation theory here, how often would you expect an expert to say, "I am not an expert in this"? It would be rare.

However, if we do find someone whose expertise is fully applicable to the situation in question, then their needs will not be counter to our own. In other words, if someone is a verifiable expert on the issue at hand, and we all recognize their range of expertise, then they can achieve esteem and self-actualization by applying it. Therefore, it is incumbent on those who are responsible for making decisions to ensure that they have the right experts. They should, first, identify the real problem (such as data analysis instead of medicine). Then, second, clarify what they are looking for in an expert. To understand what makes an expert, we must explore how people make decisions and form intuition.

Human Judgment

In 1954, eleven years after Maslow penned his famous paper, a 21-year-old college graduate named Danny Kahneman joined the Israeli Defense Force. This was not unusual, as every young Israeli serves at least two years in their country's military. However, having been one of the first students to graduate Hebrew University with a degree in psychology, his first major assignment was anything but standard. Danny was part of a new division, the psychology unit, and his job was to oversee the selection process for officers.

Over the next year, as he observed how the results of his selection process turned out during officer training, he tried several permutations to perfect the process. In particular, he noticed that the interview process seemed to produce virtually no consistent results in selecting successful officers. In the end, he modified the process to become more logic-based. Instead of asking the interviewer to form an opinion about the candidate, he forced them to ask questions that prompted very specific answers about behaviors and tendencies. With the answers to these questions, he utilized an algorithm that he created to determine likelihood of success. This approach not only worked for the IDF, it also predicted the success these recruits would have in almost any job.

This was the first major breakthrough in Daniel Kahneman's long career. He would go on to earn a Master's and PhD in psychology from the University of California, Berkeley, and write a number of books. His theories on human decision making would spawn the subject of BEHAVIORAL ECONOMICS, for which he would be the first psychologist to win the Nobel Prize in economics in 2002. Out of this work, several areas of our economy have been revolutionized. Michael Lewis, one of my favorite authors, has written several books on the subject.

MONEYBALL tells the story of how Billy Beane revamped the 2002 Oakland Athletics roster using data. THE BIG SHORT tells the story of how a handful of financial investors correctly predicted (and made huge sums of money betting on) the 2008/2009 housing collapse. Finally, his book THE UNDOING PROJECT tells the story of Daniel himself, and his partner Amos Tversky, and how they started the rise of statistical psychology.

Throughout all these stories, whether you are talking about interviewers looking for military officers, financial investors, or scouts for professional sports teams, the themes are the same. Our experts often don't know as much as they (or we) think they know. Why? Bias.

One of the fundamental foundations of traditional economics is that human beings are rational. This is why supply-and-demand models can effectively predict prices. If there is an excess of a product on the market and people will always (rationally) spend the least amount of money possible for a product, then prices will drop as producers compete for consumers. If human beings are not rational, and will pay more for a product despite a cheaper option, the model will not work. While these macro-economic theories generally work, behavioral economics recognizes that human beings are not always rational. In fact, there are a number of heuristics we employ to assign value or probability. A heuristic is a mental shortcut that allows an individual to make a decision, pass judgment, or solve a problem quickly and with minimal mental effort. In his book THINKING FAST AND SLOW, Kahneman explains that it is heuristics that cause irrational decisions, not emotional states as previously thought:

> *Social scientists in the 1970s broadly accepted two ideas about human nature. First, people are generally rational, and their thinking is normally sound. Second, emotions such as fear, affection, and hatred explain most of the occasions in which people depart from rationality. We documented systematic errors in the thinking of normal people, and we traced these errors to the design of the machinery of cognition rather than to the corruption of thought by emotion.[2]*

The *machinery of cognition* refers to the heuristics, the methods we form, based on experience, of arriving at conclusions easily. These heuristics form the bias with which we evaluate a situation. Over the course of the last twenty years, hundreds of research projects have examined and documented dozens of different common heuristics. In THINKING FAST AND SLOW, Kahneman explores several of them:

- *The law of small numbers*: we assume a small sample accurately represents a larger one
- *Availability*: we put more weight on information that is easily recalled or in recent memory
- *Anchoring*: we set a bias based on an initial value, regardless of applicability

For a quick read about behavioral economics and references to fascinating studies, try BEHAVIOURAL ECONOMICS – A VERY SHORT INTRODUCTION by Michelle Baddeley. And yes, that is the British spelling of behavioral.

---

[2] D. Kahneman, THINKING, FAST AND SLOW (New York: Farrar, Straus, and Giroux, 2011), 8.

- *Confidence*: we place more value on knowns versus unknowns
- *Representiveness*: we place a bias on correlations between facts that may have no relationship

There are dozens more, but one that is particularly applicable here is *Availability*. Kahneman explains an example like this:

> *Students of policy have noted that the availability heuristic helps explain why some issues are highly salient in the public's mind while others are neglected. People tend to assess the relative importance of issues by the ease with which they are retrieved from memory—and this is largely determined by the extent of coverage in the media.*[3]

If one were to theorize as to why COVID-19, while certainly a real threat, seemed to disproportionately outweigh other societal problems, the *Availability* heuristic helps explain. It can also explain why an expert in a plant may place an artificially high probability of failure on an asset if he or she witnessed a recent failure of a similar asset. Or why a parent may impose out-of-the-norm restrictions on kids going to a friend's house after reading a story about abduction. It is not that these scenarios indicated a shift in the probability, but our perceived risk was higher due to the availability of information in our mind.

So that's it? Human judgment can't be trusted, therefore, experts are out? No. In fact, Kahneman explains this too. In fact, he worked with Gary Klein, a leader in a different focus area when it comes to studying human decision making. Naturalistic Decision Making (NDM), Klein's area of research, is based on experts acquiring an "intuition" based on their experience. NDM also asserts that this intuition is very valuable, and often a good reference point for decisions. Together, Kahneman and Klein worked to answer this question: *When can you trust a self-confident professional who claims to have an intuition?*

---

[3] Kahneman, THINKING, FAST AND SLOW.

Their conclusion? "Do not trust anyone, including yourself, to tell you how much you should trust their judgment." Instead, they said there are "two basic conditions for acquiring a skill."

- An environment that is sufficiently regular to be predictable
- An opportunity to learn these regularities through prolonged practice

"When both of these conditions are satisfied," Kahneman writes, "intuitions are likely to be skilled."[4]

With this "test" we can examine when to accept someone's judgment. In the case of the coronavirus pandemic, there were no doctors who could meet these criteria. As such, we should not have based our decisions solely on their intuition without additional data to back up their judgment.

Losing Experts

As we explored in Chapter 2, experts made tremendous advancements in major process industries, like refineries, from the 1980s to 2010. Improved efficiencies, lower costs, and better reliability all came from experts. However, during that period, experts were solving different problems than we face today. A majority of those issues were isolated to a specific area and had occurred more frequently. Hence, the experts met the criteria listed above. For example, if one pump failed eight times between 1991 and 2002, and the machinist or maintenance engineer examined the pump in each of those instances, he would be an *expert* in that pump's failures. He could assess the needs of the pump and designate the best improvement to keep the pump running longer.

Today, we are struggling to make these kinds of improvements because we are losing experts who meet Kahneman and Klein's criteria. This is caused by three things.

---

[4] Kahneman, THINKING, FAST AND SLOW, 240.

1. Problems or opportunities to improve are too infrequent for an individual to gain enough experience to be an expert.

2. People are not staying in a position long enough to gain enough experience.

3. Modern-day knowledge workers operate in a much higher degree of distraction, which prevents the study of key issues and variables to gain expertise.

Today, if a pump fails twice in an eleven-year period, but the failures are linked to different causes that involved other equipment items, the machinist or maintenance engineer won't meet the expert criteria for that pump. The plant personnel must turn to more sophisticated use of data, from multiple sources, to improve the design and reliability of the unit. It is much harder to become an expert in these types of problems.

We have another problem across virtually all industries: *turnover*. It's hard to find reliable information on turnover rates prior to 2000, but most companies publicly recognize the increase. Anecdotally, oil and gas companies, when I entered the industry in 1998, fully expected their employees to stay through retirement. Some estimates are that turnover at that time was below 10 percent. For the past twenty years, the rate has ebbed and flowed, but the numbers are staggering. In 2019, the average turnover rate in the United States was over 40 percent.[5] People are changing jobs, on average, every two to three years. For those under thirty, it is nearly once per year, on average. The result is that most people are not staying in a job long enough to become an expert.

Finally, regardless of time spent on a job, our work practices today are far different than they were forty years ago. In his brilliant book, DEEP WORK, Cal Newport heralds the practice of digging in deep to focus, study, and create.

---

[5] "Job Openings and Labor Turnover Survey News Release," U. S. Bureau of Labor Statistics, last updated March 17, 2020, https://www.bls.gov/news.release/archives/jolts_03172020.htm.

Deep work is necessary to wring every last drop of value out of your current intellectual capacity. We now know from decades of research in both psychology and neuroscience that the state of mental strain that accompanies deep work is also necessary to improve your abilities.[6]

Conversely, he laments the fact that most employees today, especially the knowledge workers, rarely have time for this. In a world where we are constantly overloaded with meetings, emails, office pop-ins, text messages, phone calls, and electronics, almost no one is taking the time to study an issue in detail, research information, and form new or consequential ideas.

Big trends in business today actively decrease people's ability to perform deep work, even though the benefits promised by these trends (e.g., increased serendipity, faster responses to requests, and more exposure) are arguably dwarfed by the benefits that flow from a commitment to deep work (e.g., the ability to learn hard things fast and produce at an elite level).[7]

In short, the vast majority of us are not working in the right ways to become an expert. So while our problems have become more complex, there are fewer and fewer experts, and they aren't experts in the right things. That is why we need data.

But data, of course, has its own problems.

---

[6] C. Newport, Deep Work (New York: Grand Central Publishing, 2016), 3.
[7] Newport, Deep Work, 52.

Chapter 6

The Data Problem:
Why We Use Experts

For the last ten years, data has been all the rage. Whatever you call it—*big data, machine learning, artificial intelligence,* the *internet of things*—if you want investors, do something with data. All of these labels refer to a simple idea: that handling and analyzing data can bring a whole new world of solutions to problems and produce a whole new world of value. Job openings are all over the place for data scientists or data analysts. Universities have been establishing institutes and consortiums and created degree programs linked to data science. In October 2019, presidential candidate Andrew Yang proclaimed during a debate, "Right now, our data is worth more than oil."

It seems that everyone is trying to jump on the data train, from coffee growers, to healthcare, to oil and gas, to home air conditioning systems. Yet for all the investments, the startups, and the hype, results have been mixed. Sure, we all see the results of rapid data analysis when we log onto Amazon and it knows what we want to look for before we even type it. Google routes us in all sorts of directions based on traffic flow and user input. We see fantastic results in image and video recognition. We hear about new algorithms that are helping investment advisors pick better stocks for their clients and wonder if our guys are using that too.

At the same time, there have been hundreds of stories of initiatives started and stopped at large companies trying to leverage data. Startups that got venture capital funding are struggling to sell their wares. Even companies that have found some success in getting new data analytics going are finding it much harder than expected to quantify the value. One oil executive told me, "We've seen two dozen different technology and consulting companies come in here and pitch us on their data platform, but when we start pushing them on how they will make us perform better, the answers are empty. It seems they want us to just plug into their computers and see what happens." Hardly a recipe for massive investment, even before the coronavirus depression.

If it seems like something is missing, it is. While our computer and software systems are bigger and faster than ever, the ability for artificial intelligence to magically find new business solutions is not realistic. What is realistic, given enough inputs, is that a software system can identify patterns and anomalies that a person can't see fast enough, if at all. But all this logic is dependent on data. And not just a lot of data, but the *right* data. Right now, most data science projects are suffering from one or more of three big hurdles—insufficient data, inaccurate data, or the wrong data.

Insufficient Data

If I held up a coin you had never seen and asked you, "What is the probability that it lands on heads?" You would say, "50 percent." No testing, no experience, just intuition. Now, let's suppose I hold this up, a ten-sided die.

This time, I ask, "What are the odds that, if rolled, it comes up one, two, or a three?" You would answer, "30 percent." Again, almost no thought. You know the answer. Intuition. You have enough experience with simple probabilities like this and can run a simple fractional probability in your head without work. According to Daniel Kahnemen, you are thinking fast.

Now, suppose we wanted a computer to calculate the odds of rolling a one, two, or a three. What would we do? We could enter in the total number of sides and the number of sides with a one, two, or a three on it, but that means that we are doing the work, and the system is just a calculator. How would we do it? The statistical approach would be to roll the die over and over, enter the results into the computer, and let it determine the probability *empirically.* Let's see how this works.

I went into Microsoft Excel and created a one hundred-cell sheet. I entered the formula for random number generation, with a range of one to ten, simulating a roll of the dice one hundred times. This is what I got:

3	2	8	10	8	10	3	3	9	2
7	9	6	8	10	4	4	6	4	10
4	1	1	10	10	1	5	2	10	1
7	2	5	5	4	7	9	5	2	4
9	7	2	9	5	4	2	10	8	7
8	4	9	4	6	4	8	8	7	5
7	5	6	1	1	10	7	10	4	7
1	7	1	7	2	5	7	4	1	9
3	5	8	2	10	1	4	3	10	3

Ten rolls came up one, nine rolls came up two, and six rolls came up three. This totals twenty-five, so the computer assigns a 25 percent chance this dice will come up with a one, two, or a three. Now, we may quickly say there just aren't enough sample points. So, for kicks, I tried it with 1,000 cells: 106 rolls came up one, 104 rolls came up two, and 109 rolls came up three, for a total probability of 32 percent. Much closer to 30 percent than 25, but still off. The data is not altogether worthless, but this example explains the challenge. One hundred samples are not a lot of die rolls, healthcare visits, or purchases on Amazon, but it is too many refinery failures, airline crashes, or economic recessions. The fact is, before we start identifying patterns of data for events, we have to make sure we have enough events. And for many big events, like a global pandemic, we simply don't have that.

Even where we should have enough data, there are barriers. Thousands of people a day walk into the doctor's office around the country. Loads of data are gathered, but we can't access it. Why? HIPAA laws. Fear of big brother using your data against you or companies refusing to insure you because they find out you have a preexisting condition has locked down data so that no one can really use it. Also, inside big industrial facilities, various systems have been set up over the years by different departments, and none of those systems talk to each other very easily. In some cases, those systems are proprietary, closed systems that cannot be easily accessed. In other cases, some of those systems are designed not to talk, as the owners hope that isolating critical systems will lower the risk of cyber-attacks. Several of the world's largest companies have announced multi-billion-dollar initiatives designed to do one thing, give them easier and more organized access to the data they already own.

Inaccurate Data

"Garbage in equals garbage out." That's how the saying goes. I have heard it repeated hundreds of times in my career as people work to get their data in "good shape" before working on more advanced data management programs.

While this may be a bit misdirected (see the next section), this is probably the easiest problem to solve. For decades, we have been focused so heavily on gathering data, we didn't focus on making sure it was right. This has been true for virtually every industry—automotive, healthcare, and oil. Think about how many times you have gone into the doctor and they have quickly checked your vitals, including blood pressure. I have been in, had my blood pressure checked, had them tell me it was 138 over 85, then stopped at a CVS on the way home and gotten a reading of 118 over 70. Which was right? Perhaps both were right at the time. Perhaps my blood pressure was higher because I was stressed before having my blood drawn. Regardless, no one at the doctor's office asked any questions. What is the point of checking it if you don't make sure it is meaningful? If you are only checking for problems in the moment, fine. But this lack of information eliminates the ability to look at trends, which is where the real opportunities lie.

As we discussed in Chapter 4, the oil industry began taking UT thickness readings on pipe in the 70s, and by the 90s, plants had spent millions of dollars to set up very basic data gathering programs. Since this had become a new requirement in API 570 in 1993, almost all the attention was on *getting* the data, not making sure it was right or that it could be used effectively. Therefore, the value coming out of these efforts was miniscule relative to what it could have been. Having recognized this, most plant reliability or mechanical integrity managers have tried to push their people to focus much more heavily on getting and using accurate data instead of just getting data. Since it is hard to change decades of habits, some facilities have been through two or three iterations of data setup.

In the end, this is about mindset. In UPTIME and MAINTENANCE AND RELIABILITY BEST PRACTICES, Campbell and Reyes-Picknell, talking about computerized maintenance systems, say, "All of these software tools rely heavily on accurate input of the right data." However, they go on to say, "The right data include a description of the as-found state of the asset at the time of a repair or preventive renewal."[1] In other words, they combine data and information, which are not the same. Gulati has a section on Data Collection and Data Quality, and says, "Data is the key ingredient in performance management."[2] Both books focus heavily on leadership and management. Gulati talks about a "culture of reliability," and Campbell and Reyes-Picknell have an entire chapter about people.

Quality data is all about leadership, management, culture, and people. As leaders, we must make sure that our teams understand that it isn't just about gathering data; it's about what we do with it. Good quality *and* good analysis are crucial to good conclusions. In fact, making sure our data is right can be less important than making sure we have the right data.

The Wrong Data

For anyone thinking, "Come on, Sitton. People were looking at a *lot* of data around coronavirus." On March 25, most schools had closed around the country, and within the next two days, half of US states had issued stay at home orders.[3] The World Health Organization issued daily status reports. The following is some of the information published that day.[4]

---

[1] J. V. Reyes-Picknell, UPTIME, 185.

[2] Gulati, MAINTENANCE AND RELIABILITY, 297.

[3] U.S. State and local government responses to the COVID-19 pandemic," Bomis, updated July 8, 2020, https://en.wikipedia.org/wiki/U.S._state_and_local_government_response_to_the_COVID-19_pandemic.

[4] "Coronavirus disease 2019 (COVID-19) Situation Report- 65," World Health Organization, United Nations, updated March 25, 2020, https://www.who.int/docs/default-source/coronaviruse/situation-reports/20200325-sitrep-65-covid-19.pdf?sfvrsn=ce13061b_2.

Table 2 – WHO COVID-19 Daily Status Report Data: March 25, 2020

SITUATION IN NUMBERS Total (and new) cases in last 24 hours Globally 413,467 confirmed (40,712) 18,433 deaths (2,202)	
Western Pacific Region 97,766 confirmed (1,186) 3,518 deaths (16) European Region 220,516 confirmed (25,007) 11,986 deaths (1,797) South-East Asia Region 2,344 confirmed (354) 72 deaths (7)	Eastern Mediterranean Region 29,631 confirmed (2,416) 2,008 deaths (131) Region of the Americas 60,834 confirmed (11,390) 813 deaths (248) African Region 1,664 confirmed (359) 29 deaths (3)

WHO Risk Assessment: Global Level – Very High

On page three of the report, it showed Italy:
Total Cases: 69,176 (5,249 new in the last 24 hours)
Total Deaths: 6,820 (743 new in the last 24 hours)

And on page four, the United States had:
Total Cases: 51,914 cases (9,750 new in the last 24 hours)
Total Deaths: 673 (202 new in the last 24 hours)

At this point, Italy had a 9 percent mortality date, and the climb in US deaths was alarming. Thirty percent of our deaths had occurred in the past twenty-four hours. What if our mortality rate was like Italy's? The day before, Governor Cuomo in New York predicted that the city may need 140,000 beds versus the 53,000 they actually had. Finally, the front-page story in several newspapers over the previous week was that some models being used by the CDC were showing that as many as 1.7 million people could die from this. With the recent uptick in the death rate, that wasn't too hard to believe.

That was it. The data was in. Obviously, we had to lock down the country. I will admit, my wife and my father were ready to lock themselves in and board up the doors for the long haul, regardless of what governments did. But at some point, some questions should have been asked. *Why are some countries' mortality rates so low? If the disease can take up to fourteen days to incubate, how could it have spread so far and so fast in thirty days, to trigger a WHO alert? Is everyone at the same risk of getting this? If risks are this high, what will lockdowns do?*

Two pieces of data were incongruous. On March 25, it seemed like coronavirus could spread overnight, with thousands of new cases being reported every day. Yet contact tracing showed much longer periods for contagion. So, how could that be? The answer is, it couldn't.

The math indicates that the disease had to be circulating long before February. Given this quandary, a group of researchers at Stanford began to dig in. Within two weeks of the March 25 lockdowns they a published the study.[5] They found "The reported number of confirmed positive cases in the country on April 1 was 956, 50 to 85-fold lower than the number of infections predicted by this study." In other words, the only conclusion to draw was that the disease had been circulating in the US much earlier than reported, and most people had no symptoms, which is why we didn't notice. This would be confirmed by the State of California, when they reported on April 26 that two deaths occurred on February 6 and 17, meaning that coronavirus was circulating at least weeks earlier, probably longer. If their low end is right, and fifty times as many people actually had the virus, then our mortality rates were not 2 percent, they were more like .04 percent.

What about risks on a demographic basis? It is difficult to understand how this seemed to get overlooked so easily. In short, we all knew that two factors seemed to weigh heavily on

---

[5] "Covid-19 Antibody Seroprevalence in Santa Clara County, California," MedRxiv, Chan Zuckberg Initiative, updated April 17, 2020, https://www.medrxiv.org/content/10.1101/2020.04.14.20062463v1.full.pdf.

the risk associated with COVID-19 – age and preexisting conditions. Unfortunately, it is hard to go back now and get slices of data on certain days, but we can get close. On April 3, the CDC published its Morbidity and Mortality Weekly Report (MMWR) for the prior week, up to March 28.[6] In that report, it showed the following table:

Table 3 – Hospitalization and ICU Rates for COVID-19 Based on Preexisting Conditions

| Age group (yrs) | Hospitalized without ICU admission, No. (% range†) | | ICU admission, No. (% range†) | |
| | Underlying condition present/reported§ | | Underlying condition present/reported§ | |
	Yes	No	Yes	No
19-64	285 (18.1 - 19.9)	197 (6.2 - 6.7)	134 (8.5 - 9.4)	58 (1.8 - 2.0)
≥65	425 (41.7 - 44.5)	58 (16.8 - 18.3)	212 (20.8 - 22.2)	20 (5.8 - 6.3)
Total ≥19	710 (27.3 - 29.8)	255 (7.2 - 7.8)	346 (13.3 - 14.5)	78 (2.2 - 2.4)

What this data tells us is that people from age 19-64 who contracted COVID-19 without preexisting conditions were admitted to the ICU at a 20 to 25 percent rate of those *with* preexisting conditions. Also, only 30 percent of people admitted to the ICU did not have preexisting conditions. Now, in order to translate that to probability by age group, we can look at total ICU admissions, regardless of preexisting conditions. On March 18, the CDC put out an early MMWR with this table:[7]

[6] "Preliminary Estimates of the Prevalence of Selected Underlying health Conditions Among Patients with Coronavirus Disease 2019- United States, February 12-March 28, 2020," Centers for Disease Control and Prevention, U.S. Department of Health & Human Services, updated on April 3,2020, https://www.cdc.gov/mmwr/volumes/69/wr/mm6913e2.htm?s_cid=mm6913e2_w.

[7] "Severe Outcomes Among Patients with Coronavirus Disease 2019 (COVID-19)- United States, February 12-March 16, 2020," Centers for Disease Control and Prevention, U.S. Department of Health & Human Services, updated March 26, 2020, https://www.cdc.gov/mmwr/volumes/69/wr/mm6912e2.htm.

Table 4 – Impact Rate of COVID-19 by Age Group

Age group (yrs) (no. of cases)	Hospitalization rate (%)	ICU admission rate (%)	Case-fatality rate (%)
0 - 19 (123)	1.6 - 2.5	0	0
20 - 44 (705)	14.3 - 20.8	2.0 - 4.2	0.1 - 0.2
45 - 54 (429)	21.2 - 28.3	5.4 - 10.4	0.5 - 0.8
55 - 64 (429)	20.5 - 30.1	4.7 - 11.2	1.4 - 2.6
65 - 74 (409)	28.6 - 43.5	8.1 - 18.8	2.7 - 4.9
75 - 84 (210)	30.5 - 58.7	10.5 - 31.0	4.3 - 10.5
≥85 (144)	31.3 - 70.3	6.3 - 29.0	10.4 - 27.3
Total (2,449)	20.7 - 31.4	4.9 - 11.5	1.8 - 3.4

This table tells us that for those age 54 and under, there is a .2 to .4 percent chance of fatality. When we integrate the data from the previous table that informs as to the different ICU rates with and without preexisting conditions, we find that someone age 54 and younger who has tested positive for coronavirus without preexisting conditions has a .07 to .13 percent chance of dying.

Finally, let's assume that the Stanford study is on the right path, and that the incident rate for coronavirus has been woefully underreported because of a lack of symptoms and a lack of testing. But instead of a factor of 50 to 85, let's apply a conservative factor of 10. That would mean the chances of dying for those under 55 who have no preexisting conditions just fell to 0.007 percent to 0.013 percent. By comparison, the annual mortality rate for pneumonia (normally 50,000 people per year)[8] is 0.16 percent,[9] and the fatality rate for the flu

[8] "Top Pneumonia Facts-2019," American Thoracic Society, American Thoracic Society Foundation, https://www.thoracic.org/patients/patient-resources/resources/top-pneumonia-facts.pdf.

[9] "Influenza/Pneumonia Mortality by State," Centers for Disease Control and Prevention, U.S. Department of Health & Human Services, updated April 26, 2020, https://www.cdc.gov/nchs/pressroom/sosmap/flu_pneumonia_mortality/flu_pneumonia.htm.

for those under 50 is 0.008 percent.[10]

Let's accept, for the sake of argument, that there had not been earlier or larger untested circulation. Rather, this disease spreads so fast and with such high hospitalization rates that it was going to ripple through our country at an unparalleled rate. At some point, the data to be considered would be this: if it is so contagious, and only a few have been infected, what will happen when we open back up? In other words, as of May 15, only 1.45 million out of 330 million total Americans had been confirmed positive. This means, if the disease did not spread earlier and farther than previously thought, then 99.5 percent of our population has yet to be exposed. If that were true, and the transmission, hospitalization, and mortality rates are as high as the early models suggested, then we should have recognized that we would prevent nothing. Rather, we simply delayed things by three months by shutting down one-third of our economy. In July, as governments reopened and testing ramped up substantially, this is exactly the reality that landed. And when it did, a new wave of panic set in. At that point, governments were trying to figure out how to strike a balance. Namely, how could we protect those at high risk through a combination of specific restrictions and significant social distance guidelines but preserve some operation of life to stave off another round of economic disaster? The key question was: why didn't we do this in March?

At the time of the writing of this book, it is, admittedly, early for reflection on March, April, and May 2020. But things are changing so fast it requires that we ask ourselves, as a society, where we may have made mistakes. To do that, we must step back from the simplistic lens of "saving lives" by reducing exposure to COVID-19, and instead look back through a lens of data, risk, and overall societal cost. In doing so, it appears that we did a very poor job of pulling together a complete data set to evaluate what actions to take in response to the coronavirus pandemic. In particular, by looking at the wrong data (in this case because it was

---

[10] "Estimated Influenza Illnesses, Medical visits, Hospitalizations, and Deaths in the United States- 2017-2018 influenza season," Centers for Disease Control and Prevention, U.S. Departments of Health & Human Services, updated November 22, 2019, https://www.cdc.gov/flu/about/burden/2017-2018.htm.

an incomplete set), it appears that we drew two erroneous conclusions, and completely missed a third. First, the set of data we
considered led to vastly overestimated perceptions of risk to our
population. Second, it appears that we overestimated the long-
term positive impact of government-imposed lockdowns. Third,
it appears that no one fully analyzed, or understood, the massive
negative impacts of those lockdowns on the rest of our society.

This might be the most powerful example in history of how
looking at the wrong (or incomplete) set of data can lead to inaccurate analysis, poor decisions, and tragic consequences. The
key here is that it wasn't because the data wasn't available. It was
because the people making decisions didn't know what to do with
it, and hence drew bad conclusions from simple correlations.

House of Cards

In April 2019, Kalev Leetaru, an experienced software entrepre-
neur who writes about AI and Big Data for Forbes Magazine,
wrote an article entitled "Our Entire AI Revolution Is Built On
A Correlation House Of Cards" that appeared on Forbes.com.
In it, he railed on the dependence on correlation of most data
initiatives.

> *We are literally betting the future of our planet on statisti-
> cal correlations. Today's deep learning systems don't actu-
> ally "understand" the world. They do not take their reams of
> inputs and abstract upwards to high-order entities described
> by properties and connected to other entities through relation-
> ships and transitive causation.*
>
> *Instead, our most powerful AI systems merely discover
> obscure patterns in vast reams of numbers that may have
> absolutely nothing to do with the phenomena they are sup-
> posed to be measuring. Most importantly, we have little way
> of testing whether those correlations are wrong until our algo-
> rithms fail in the most spectacular, and unfortunately some-
> times fatal, ways.*

The early era of AI was built on the idea of machines understanding the world in the same way we do: as semantic abstractions and relationships whose interactions are guided through causation. Building such AI systems is really hard, so we've gone back to what's easy: building correlation engines. It is a lot easier to build a machine that can spot patterns in a pile of numbers than it is to build a machine that can take those numbers and use them to build a mental model of the world they describe.

Putting this all together, we have built our modern AI world upon a correlation house of cards that is beginning to buckle under the strain of all our hopes and aspirations colliding with statistical reality.[11]

In short, we don't have the right data for our simple systems, and we don't have the right systems for our simple data. Leaders from business to politics lack confidence in data and systems that are possibly inadequate, inaccurate, or incomplete. We are hesitant to put our trust in unproven, impersonal machines during stressful moments. As a result, we call on experts to help us make crucial decisions. Sure, we may be making the wrong call sometimes, but we feel that we are making the best call we can, with the people we have.

But this "house of cards" is where opportunity lies. If we want to use data to augment our experts and use experts to augment our data, we must build the machines that can analyze the "abstractions and relationships whose interactions are guided through causation." The fact is, it will take both data and experts.

And that is where our solution begins.

---

[11] "Our Entire AI Revolution Is Built On A Correlation House Of Cards," Forbes, Forbes Media LLC, updated April 20, 2019, https://www.forbes.com/sites/kalevleetaru/2019/04/20/our-entire-ai-revolution-is-built-on-a-correlation-house-of-cards/#1125f6034969.

Chapter 7

The Solution:
The New Model of Data and Experts

"Iᴏ ᴀ ʜᴜᴍᴀɴ ʙᴇɪɴɢ ᴄᴀɴ sᴇɴsᴇ ɪᴛ, a human being can quantify it. If he can quantify it, he can learn about it."[1] This was the philosophy of Sid Mejdal, the chief data analyst for the Houston Astros, when he transformed the qualitative opinions of scouts into numbers and measures. While Sid was pushing the envelope in baseball, experts across multiple facets of our society were doing the same. In Fᴀɪʟᴜʀᴇ ᴏꜰ Rɪsᴋ Mᴀɴᴀɢᴇᴍᴇɴᴛ, Hubbard continually cautions the reader about the pitfalls of qualitative measures such as scoring systems and subjective assessments of probability. Historically, the biggest failures in decision making were not due to bad judgment; they were due to an inaccurate assessment of the situation. These inaccurate assessments are enabled, almost entirely, by acceptance of qualitative judgment. Our solution starts there, by limiting that practice.

Now, this is a bold statement. Are you telling me that a father should stop and do quantitative risk analysis before telling his sixteen-year-old daughter she cannot go on a date with the twenty-one-year-old bartender? Or that we should stop and calculate the odds of staying safe while stopping at a shady gas station for a bottle of water in the middle of the night? Or that we need to quantify the probabilities of success in buying our spouses the right birthday present? No. As we discussed in Chapter 5, our intuition has formed over repeated experiences, and should be

---

[1] Ben Reiter, Asᴛʀᴏʙᴀʟʟ, 50.

trusted. Also, our solution is *not* intended to address the singular decisions with one set of drivers and outcomes—which car should I buy, should I invest in this stock, etc.

While these theories could be applied to any risk-reliability analysis, the focus of this book has been on those decisions—dozens or hundreds or thousands of them—required to resolve complex problems with complex systems. *How do we assemble and manage the best possible baseball team, considering over 250 minor and major league players? How can we develop the best reliability program for thousands of assets in a refinery? How do we develop an operating and improvement plan for a sophisticated water treatment facility? How can we attract, hire, and integrate hundreds of new employees into a growing company? Or, how should we handle a virus that is spreading more rapidly than any in a hundred years and striking fear around the world, with different environments, risks, and resources in each of the country's 3,141 counties?*

These complex problems and systems require us to think differently about how we make decisions. Not only must we integrate the resources at our disposal more effectively, we also need to shift our mindset away from making thousands of individual decisions, and instead shift to integrating all those decisions into a defined method. I will refer to this as the quantitative decision process.

Quantitative Decision Process

If you were to google "decision process," you would likely find explanations and images laying out something like this:

This process, or some version of it, has been used for decades to ensure that groups of people took the time to consider and discuss all options before making decisions. The main thrust of this model is to ensure that knowledge and people are leveraged. I have used something like this myself with great success when making difficult choices in business. However, when faced with more broad, systematic problems and complex systems, there are too many variables, so this process comes up short.

In a quantitative decision process, we must rethink our roles. Since a quantitative model *assumes* that the decision will be based on the data, instead of following the flow of people, we monitor the flow of that data. Here is the simple model:

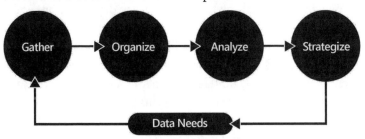

Figure 5 – Quantitative Decision Process

While somewhat elementary, this diagram is crucial to understanding how we must shift from what we are doing today. At first, it may appear that this approach is much like our current practice. However, when presented with a complex challenge in today's world, we tend to focus most (if not all) of our energy into two of the four areas represented by circles, which is why we struggle to keep our decision processes quantitative and data focused. To explain, we need to ensure that we understand these four areas.

Gathering data is the process of obtaining raw measurements and capturing them for future use. This should be thought of as the point at which our world as decision makers touches the physical world of system activity. This could be measuring thickness, temperature, or vibration of a machine. It could be measuring number of cars, number of accidents, and number of injuries on a road. It could be testing a new employee with a DISC or Myers-Briggs profile assessment. Or it could be acquiring the age, movement patterns, and preexisting conditions of a coronavirus patient. In each of these cases, the four key parts of gathering data are capturing the data, ensuring the data is accurate (quality), identifying the correct quantification (if needed), and depositing the data in the appropriate system to be reviewed and applied later.

Organizing data is more than simply sticking it into its place on a spreadsheet. Data is only useful if it is formatted, tagged, and arranged into areas that can be easily scanned or used in later analysis. For example, when people are trying to lose weight, some will start weighing themselves every day. There are scales that will take that data and store it in an application on your phone for easy trending. As you would expect, a 2018 study by the American Heart Association found that people who track their weight regularly do tend to be better about losing it.[2] However, the most likely driver behind this is not that the data tells you about how to lose weight, but rather, that the people who focused on it regularly tended to make better lifestyle choices. If that data were organized, relative to the last meals eaten, exercise activities, sleep, and other life events, it would be much more instructive.

---

[2] "Abstract 10962: Temporal Patterns of Self-Weighing Behavior and Weight Loss in the Health eHeart Study," Circulation, American Heart Association, updated November 5, 2018, https://www.ahajournals.org/doi/10.1161/circ.138.suppl_1.10962.

Yes, part of this is gathering the right data (meals, sleep, exercise, number of steps in a day), but organizing it in ways that show its relevance is just as important. In fact, it is often in this organization and analysis that we recognize the other data that is needed. This missing data is either filled in with assumptions or left as unknowns.

Analyzing data may seem obvious, in that someone must go through and review all the data and identify patterns and trends. This can be done through manual calculations or estimation, or through hard formulas and computers. In other words, both "expert" analysis and algorithms are included in this step. Both are crucial. In fact, in the early phases, analysis usually begins with data analysts and experts in the field at hand working together to "crunch the numbers." But in today's world, this must quickly evolve. Either we must develop the more sophisticated algorithms to derive more insightful results, or we must pull enough data with enough organization to allow existing machine learning algorithms to run and identify correlations in our data.

One key here is often missed. We must attempt to perform much of the data analysis before we know what answer or solution we are looking for. Why? Data is objective. Analysis sometimes is not. It is often tempting to start reviewing data expecting a certain result, and then find the analysis that fits. Do not ask, for example, "Will staying home save lives?" Instead, ask for a specific quantifiable answer. Something like, "What is the total risk to people if we do nothing, compared to the total risk to people if we shut down the country?" As analysis is taking place, strong teams will recognize where they do not have the experts or the algorithms, and simply lack the ability.

Strategizing is the step where the decision makers, independent of data gathering, organization, and analysis, get involved. Their role is three-fold. First, they should test the process and the results by comparing the results against empirical situations. If certain events have driven calculated risk abnormally high, does this match other experiences when abnormal events have occurred? Second, they must assess the team to ensure the analysis was comprehensive, i.e., confirm that the experts involved were qualified (passed the Kahneman-Klein test in Chapter 5) to perform the analysis at hand. Third, they must look back at the data and the assumptions to see if an appropriate level of uncertainty was assigned. If these three things are good, a strategy can now be devised to make modifications to the system design, its management, or its monitoring.

The last step is identifying future *Data Needs*. Almost every plan will include steps to reduce risk and, as we discussed in Chapter 4, that means reducing uncertainty. This can be done by reducing assumptions, reducing unknowns, minimizing ability gaps, or developing algorithms to quantify expert input. All of it should reduce uncertainty. Therefore, any decisions that lead to the long-term plan should include data to be gathered and processed in the future, which can be less or more than data gathered in the past. Also, the empirical validation in the *strategize* step should integrate heavily here, as data analysis is continuously validated against empirical results.

So, which two steps tend to get all or most of our focus? While each of them is equally important, when a challenge arises, often we get all the data we can, then get our team together to start trying to make a decision on our path forward. In other words, we are always *data gathering* and when needed we jump in to *strategize*. And while some data *organization* and *analysis* is done, it is almost exclusively performed in the middle of decision making. As a result, we introduce bias as we bring together a group to create the plan. For most companies and groups, the big opportunities lie in *data organization* and *data analysis*.

Data Organization, Analysis, and Uncertainty

Organization and analysis are nothing new. However, in the context of data management, they are new. Or to be more clear, they are a new combination of skills and resources. Data organization consists of three parts: setting up the *real data*, making *assumptions,* and identifying *unknowns.* Data analysis is also split into three parts: *algorithms, expert opinion,* and identifying *ability gaps.* To help demonstrate the concept, Figure 6 provides an expanded view of the Quantitative Decision Process with those pieces broken out.

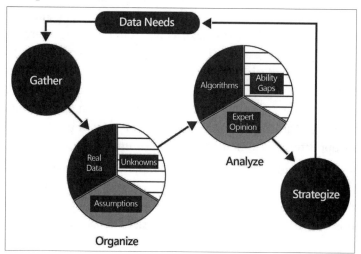

Figure 6 – Quantitative Decision Process Detail

- *Raw Data* – data in its simple numerical form, captured from field measurement or other data capture device.

- *Real Data* – data that has been moved into a logical order and structure, ready for analysis.

- *Assumptions* – data that is required for analysis, not able to be captured, but reasonably estimated.

- *Unknowns* – data that cannot (or has not) been measured, and for which there is not a logical or conservative estimate.

- *Algorithms* – the systematic calculations that turn data into insights; calculations that lead to risk, probability, and projections.

- *Experts* – the people who have the real combination of experience and information to add an intuitive analysis to the data at hand.

- *Ability Gaps* – the areas we do not know how to calculate. These can often be minimized with more expertise or more complete algorithms, but usually there is some – however small – gap.

As you can see, the elements of data organization and analysis are familiar, as is most of this discussion around a quantitative decision process. The key change from our current process is to stop focusing on the mindset of people and trying to force data into their decisions. Instead, we will focus on building processes to manage data, and apply people where appropriate. If we isolate the steps of data gathering, organization, analysis, and strategy, we introduce the least possible amount of bias, and create opportunities to make step-change improvements.

So, consider that, once data is gathered, someone needs to work through the organization and analysis of this data *independent* of the person trying to formulate the long-term plans. This does not mean people cannot work together, but their work should not be done during a combined exercise. It is simply too easy for things to get missed or for one step to artificially influence the process of another. One of the areas that can easily be missed is the identification of areas of uncertainty.

If risk is directly driven by uncertainty, then our job, especially in a more quantitative environment, is to identify and reduce that uncertainty. In Figure 6, we explored the areas of data organization and data analysis. Organization includes real data, assumptions, and unknowns. Analysis includes algorithms, experts, and ability gaps. You will note that the areas shaded in black indicate a level of quantitative activity. In other words, across the entire process flow, raw data gathering, real data organization, and algorithmic data analysis are quantified. The areas shown in grey are subjective, and the hashed areas are areas of uncertainty. Therefore, as we work this cycle, we should consistently strive to identify the areas of uncertainty or subjectivity and minimize them.

At this point, you may be scratching your head about the simplicity of this flow. After all, can it really be as simple as gather, organize, analyze, and strategize? The answer is NO. Some degrees of this are done virtually every time we have a problem. Why do some decision processes work, but others fail? It may be incomplete or improperly managed data. It could be faulty analysis or the wrong algorithms. But often the process fails before it begins. In order to effectively work through a decision process, a clear objective is needed to set the direction and measure success. If we are operating with an unclear or incorrect objective, there is no barometer against which the process can be checked. This can lead to improper application of analyses, disregarding important data, or leveraging the wrong experts. To ensure success of a decision process, we must make sure that we understand the true objective before we begin.

Objectives and Options

Sustainability, transparency, more reliable, diversity, or how about *inequality, at-risk people, disenfranchised,* or *unsafe.* What do these have in common? They are all phrases thrown around that sound good or bad and have been used to drive action in business or public policy. During the coronavirus pandemic, it was, "This is about saving lives." All sound like good things to strive for, or bad things to avoid, but they have one thing in common. They are worthless objectives. Why? There is no way to measure success, and no way to identify the cost to get there.

If there is one key area where people must drive the process of using data to assess a situation and make plans, it is in the *clear establishment of objectives* and *clear identification of options.* A terrible objective is one that sounds good, but means different things for different people. If a plant manager says, "Our top priority is safety," that's a great thing. Many industrial companies live by this. And that's okay, because it is a philosophical value, not an objective. However, if a CEO says that his company's top priority is the safety of its people, and a plant reliability manager says, "We are simply working to be the safest that we can be," then one could decide to shut the plant down to eliminate all risk. Yet, if these plants are statistically safer than driving a car, does this make sense? No. It is statistically impossible to make anything 100 percent safe, so this is not a good objective. However, "We want to make sure that our employees are 50 percent less likely to be injured at work than they are at home," is a specific, realistic objective that quantitative analysis can address. Here are some examples of good and bad objectives that have a big impact on the effectiveness of the quantitative decision process:

Bad	Good
"We want our teams to be better"	"We want to boost productivity by 10%"
"Everyone should have healthcare"	"Reduce the impact of poor health (improve quality of life) on the average American by 20%"
"We want to return value to shareholders"	"We want to ensure a 2-year return on investment for all capital expenditures"
"We need to reduce income inequality"	"We want to ensure that people with the same qualifications in the same job are getting paid a wage within 10% of each other"
"We want to become more reliable"	"We want to boost throughput by 2% but with no additional cost"
"We need to save lives"	"We want to reduce the risk (probability and consequence) associated with COVID-19 by 75%"

In this list, I intentionally used three examples from the political world: income inequality, healthcare, and coronavirus. In the world of politics and public policy, people consistently lament the fact that nothing gets done in Washington. This is partly because there are just a lot of bad ideas. But even where there are good ideas, thoughtful discussion is often precluded by simplistic rallying cries that leave no room for an actual objective. Saying that we should have free college is an objective, but a terrible one. Why? Because people don't want free college. What they really want is a good career where they feel like they are adding value to the world. Many people assume that college is the best way to achieve that, and they have been spending big dollars to get it. Is the assumption true? Does a college education really translate into a more fulfilling career? If so, is it worth the money we spend on it? Are there other ways to take 4.5 years of our lives and $90,000 to start a career that may have a higher probability of success? *That* would be a good data exercise.

Once the objective is set, there is only one other missing piece, and this is a key role for people. Options. With thousands of system components, and thousands of potential issues, we need to identify ways to make improvements. People have the ability to identify new or innovate solutions and quantify the cost and expected impact of those solutions. These solutions can be inputs into a decision process, and used in the quantified analysis. This may sound arduous. After all, for complex systems, there may be a wide range of options, but they often fit into a few buckets. For example, if you want to improve the reliability of an old car, you could do hundreds of different things. With all the different components of that car, it may seem like the options are limitless. But these options can be simplified. For most cars, you could do one of the following, to any number of different parts:

1. Make an improvement (capital investment) in the component itself.

2. Create a plan to perform regular maintenance on the component.

3. Add more monitoring, checking the condition of the component.

4. Add in a protection system of some sort, like an alarm, redundancy, or failure prevention (i.e., a ride flat tire) to warn or protect against component failure.

These basic four areas apply to almost any system, whether it is a collection of team members trying to win a world championship, a collection of pumps and piping in a chemical plant, or the

American people in the middle of a pandemic. Improvements, maintenance, monitoring, and protection are the options we can use to adjust the risk to an *individual* asset. People have been identifying and applying these solutions to components, assets, and systems for decades, and have had much success. But while this creativity and solution mindset is still important, as we discussed in Chapter 2, the next phase of reliability cannot be based on a bunch of individual analyses on individual components of the system.

If we want to perform an accurate assessment of the whole system, we cannot depend on qualitative judgments on an individual pipe, employee, pump, or patient. Instead, we must quantify the various methods of evaluating each point and tie them all together in one overall system analysis. It is not easy, but it is the only way to bring enough *precision* to the process, and make the decisions that don't waste millions, billions, or even trillions of dollars.

Which brings us to the mechanics of *Decision Precision*.

Chapter 8

How It Works:
Applying Decision Precision

QUANTITATIVE DECISION MAKING IS NOTHING NEW. As the first microprocessor was being developed, long before PCs showed up on everyone's desk, the first books on the subject were being published and distributed across college campuses. AN INTRODUCTION TO MANAGEMENT SCIENCE: QUANTITATIVE APPROACHES TO DECISION MAKING, by David Anderson, was first published in 1976. Applying statistics in bigger and better ways was the subject of dozens of books from the 1970s through the 1990s. Just last year, several new titles entered the space.

Gathering data and examining it to arrive at the best logic is as old as science. One quote often attributed to Plato is "A good decision is based on knowledge and not numbers," lending a certain bit of wit to the idea that the best results don't come from people with numbers, but people who know how to figure out what the numbers *tell* us.

So, if this has been going on for generations, what makes my approach different?

The field of statistics and quantified risk are not new, and these fields have been expanding for years as the capability of computers expands. However, there has been one big struggle in the past decade—driving more calculated results, as opposed to correlative ones. In other words, the Oakland Athletics realized that stolen bases and batting average were not as useful as slugging percentage and on-base percentage. However, scouts and managers were still deciding who to draft. The experts were still making the decisions; they were just making better ones using better data.

Over the past ten years, we have seen terms like *machine learning* and *AI* (artificial intelligence) get more airtime. The idea is that, instead of depending on experts alone, we build machines that can not only give us better data but can actually point out problems and potential solutions before people can. This opportunity gets bigger, the more complicated a system is, which is why I use examples of things like refineries and global pandemics. While this notion of machines making smarter choices is not new, something else is: how to integrate data and experts to leverage each to make the other more effective.

Decision Precision

As we explained in Chapter 7, the main area of risk, and the primary flaw in most decisions, is not the data or the experts, but in not knowing what data or experts you are missing. This is the *quantification of uncertainty*. As data moves through the quantitative decision process, one of the most important steps is to ensure that we have our hands around what data we *have*, versus what data we *need* (data organization) and what ability we *have* versus what ability we *need* (data analysis). These two steps are included when most companies say *data management*, and these steps are the bulk of where most companies are investing billions of dollars. While *data gathering* and *strategizing* are equally critical steps, they are already being done, and experienced teams know how to perform them with excellence.

Since uncertainty introduces risk, the objective is certainty. When it comes to data organization, access to a low degree of quality, structured data leads to uncertainty. Well-organized, comprehensive, quality data leads to certainty. As for data analysis, a complete set of quantitative tools produces high certainty, while simple estimates or unproven rules of thumb are uncertain. Bringing the two concepts together provides a strong sense for how much total certainty is in a decision. Since the term *precision* connotes the level of refinement or confidence in a measurement or calculation, we call this *decision precision*. It is represented in the following figure.

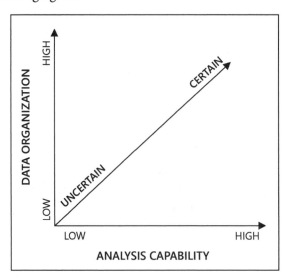

Figure 7 – Decision Precision

Inside a complex problem, if someone wants to remove uncertainty, and therefore remove risk, one must either get better data, improve the ability to analyze it, or both. The first one seems obvious. In fact, it is almost academic. "Of course, getting more data reduces uncertainty," someone could say. Not so fast. I did not say get *more* data, I said get *better* data. This is the fundamental difference between data organization and data gathering. We can always go out and get more, but if we don't get the right data, we can create more problems than we solve.

As we have covered here, decision precision in a singular judgment is neither complex nor key to the process. Questions like "Which car should I buy?" or "What college should I attend?" or "Should I take this job or stay at my current one?" or "Should we replace the water heater or repair it?" require the decision maker to evaluate whatever data they need, then make the call. However, for questions like "How much capital should we spend to improve this city's wastewater facility, what series of modifications should we make, and what is the return on that investment?" decision precision is a major ingredient in understanding how far the team, company, or organization, is ready to go.

If the precision is too low, and, therefore, the uncertainty and risk is too high, the key is not to simply do more, but to do it better. That means transitioning from less certain or subjective organization and analysis, to higher certainty, more objective functions.

From Subjective to Objective

Suppose your job requires you to do a considerable amount of speaking. For example, imagine you are a teacher, company leader, or actor, and one day you notice your voice is starting to get hoarse, especially as you try to hit higher tones. You don't think anything of it at first, but over the next few weeks, you notice that it is getting worse and worse. Finally, one day, you are singing in the shower, and your voice just gives out as you try to hit some high notes in a song that you sang easily just six months ago. Dismayed, you see your doctor.

Your primary care physician listens to your voice and notices it sounds a little raspy. She asks about the other things going on in life—exercise, diet, etc. She asks about the burning sensations in your throat and chest. You think about it and have noticed some weird sensations, especially at night when you are trying to sleep. She concludes that you probably have some sort of reflux, and it is causing your voice to get irritated. She is willing to prescribe an anti-reflux medication, but also recommends that you go see an ear, nose, and throat specialist. She refers, and you go.

At the ENT's office, you wait for two hours, and in he comes, in an obvious rush. He pulls out a very long, slender tube that holds a camera at the end. After a nurse numbs your nose, he sticks this tube up your nose, and as it travels down the back of your throat he is looking around. He has you make a couple of sounds and breathe, then pulls out the camera. "You have irritation in your throat, and a nodule on one of your vocal cords, most likely from reflux. I am going to prescribe a proton pump inhibitor, and I want to see you back here in a month." A total of ten minutes, and you're done, with a three-hundred-dollar bill and four hours of your time.

After you think about it, it seems strange. You never had reflux before and you wonder where it came from. Also, is this drug the only way to cure it? What if there is a bigger problem? You do a bit of research and find out that proton pump inhibitors are among the most commonly prescribed family of drugs in the U.S. In 2009, anti-reflux medications were the third-largest class of drug in the country with $13.6 billion in sales, representing more than 110 million prescriptions, according to IMS Health, a healthcare market research firm. They estimate that over 30 percent of the population is taking some medication for reflux, heartburn, etc. You then ask yourself, "How the heck can one-third of the population have some problem with acid reflux?"

After doing a lot more research, you find out that protein shakes, overeating, and high amounts of coffee all cause reflux, and they can compound each other. Since you are a fitness nut *and* follow an intermittent fasting diet, you are even more susceptible. After spending three hours doing research, you decide not to fill the prescription just yet, and instead make some lifestyle changes.

One week later, your reflux is gone, but your voice is still struggling. After a few more weeks, you go back to the ENT. He shoves the scope down your throat, and sure enough, the redness is gone, but the nodule is still there. He is pleased that the medication worked, until you tell him you didn't take it; then he is surprised when you tell him that you simply changed your lifestyle. Finally, when he says he should schedule you in for a surgery to remove the nodule, you politely agree, but then walk past his scheduling desk, and on to get a second opinion.

Sounds strange? This happened to me in 2019.

In the end, I had very little confidence in the precision with which my doctors were making these assessments. Why? They did not have nearly enough data. The data associated with the one symptom they could detect dominated their analysis. Any information about my lifestyle, recent changes, or correlating other symptoms was virtually ignored, because the data wasn't easily available. In addition, their analysis of the data they *did* have was based on intuition. By researching similar symptoms, treatment, and affliction rates, I could more easily diagnose what was happening by finding correlating data on the internet. It is important to note that I did not simply go onto WebMD and type in my symptoms. I looked up medical research studies and found rates of affliction, treatment, typical causes, and compared them to the changes in my own life. In other words, even though I did not have the knowledge the experts had, I made better decisions and achieved better results because I got better data and did better analysis. This example illustrates the shift in decision precision, going from subjective to quantitative.

In the data organization world, we have three buckets of needs, or inputs: *real data, assumptions*, and *unknowns*. With unknowns, we depend on users to calibrate how impactful the gaps are. This is heavily subjective and uncertain. With assumptions, we are tightening the impact, and typically base those assumptions on some history or other knowns. This is still subjective, but less so, and less uncertain. As we get into *real data*, the right structure with the right information, there is a high degree of objectivity and certainty. Moving from unknowns to assumptions to data is moving from *subjective* to *objective*, from uncertain to certain, and therefore increasing precision.

Analysis capability goes from subjective/uncertain to objective/certain in a similar way. Imagine you were plopped into the chair of a Dassault small jet and saw this:

If you have never flown a plane, then sitting in that chair, viewing LOADS of real data, you would have serious analysis gaps. All the information you need to fly the plane is there, but you don't know how to analyze and make judgments. An inexperienced pilot, one who has flown different types of planes, will have some analysis capability, but imagine an experienced crop duster pilot jumping in to fly a Boeing 747. His analysis abilities would be severely tested. Of course, someone who has years of experience in a 747 has true analysis expertise. Modern planes have sophisticated, built-in algorithms that can virtually fly those planes unmanned, all to eliminate the risks of pilot error. As we go from gaps to experts to algorithms, we are going from subjective to objective, and from uncertain to certain. Looking back at our decision precision graph in Figure 7 and applying the shifts from subjective/uncertain to objective/certain, we get the following:

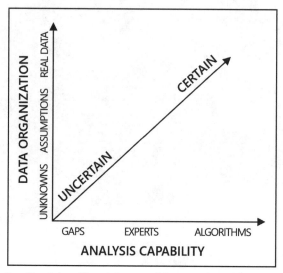

Figure 8 – Decision Precision and Increasing Objectivity

Wait, isn't it possible to have objective measures that are *uncertain?* Yes. If you think about it, what causes an objective measure to be uncertain? Every time you go to the doctor, she checks your weight, pulse, blood pressure, temperature, and respiratory rate. These are quantitative measures that we can take very accurately. However, this data can lead to uncertainty, in that your temperature could be normal, weight could go up, pulse and respiratory rate could elevate, while blood pressure is down. What the heck does that tell us? Nothing. In fact, if you had implemented a weight training routine for the last three months, you would be in better health, and yet these readings might indicate the opposite. However, the doctor will examine you and tell you that your health is fine, even improving. Is that because the objective data is less certain? No. It is because of the unknowns that are represented in the lack of data. A 2018 special edition of *Time Magazine* on "The Science of Exercise," featured an article by Dr. Jordan D. Metzl, a renowned sports-medicine physician. "If I had my way," he wrote, "medicine's four core vital signs—temperature, pulse, blood pressure, and respiration rate—would be joined by a fifth: step count."

In other words, the data we have is objective, certain, and can be used with confidence, as long as we have calibrated to what we *don't* have. That is the important part of data organization. It is also one of the crucial differentiators between traditional risk assessments or decision-making processes and reliaballing.

In considering risk of one event or one stand-alone decision, the unknowns are important, but we can be effective by quantifying our knowns and deliberating on statistical probabilities. This is how the insurance business works. The risk associated with a single driver is boiled down to average crash rates and costs given the knowns—age, education, profession, family status, etc. The unknowns become irrelevant. However, if your job is to model the overall reliability of a set of people or assets with drastically different scenarios, experiences, and unknowns, you would have to calculate the unknowns. For example, an airline company may pull one plane out for maintenance and disrupt an entire fleet. Therefore, while auto insurance companies can be confident in their risk analysis given a small set of data, the airline company cannot. The magnitude of risk in their unknowns is much higher, so they must be much more detailed in their analysis, using much higher decision precision. This focus on calibrating and addressing the unknowns is another important piece of reliaballing, not on a specific asset but on an entire complex system.

From Component to System

When looking at a set of individual assets, the typical approach people use is to try to set a limit on acceptable risk, then estimate or calculate the actual risk presented by the circumstances for each asset. For example, given an estimated number of cars driving over a stretch of road, the Department of Transportation will assign a certain level of risk for each car based on the conditions, drivers, and potential consequences on that road. If the risk is too high, they will work to lower the probability and/or consequences through design (embankment, shoulder width, guardrails, etc.) or constraints (speed limit, traffic lights, etc.).

When the performances of those assets become linked, like a series of components in an industrial process, or the population of a city or state in a viral pandemic, or a large team of people in a company or sporting event, the probabilities and consequences are linked. However, despite this very crucial difference, the same method of evaluating risk on an individual basis is the predominant method. This is not because of some faulty logic, but because, until recently, we have lacked the tools capable of handling these computations on a real-time basis.

The following image is courtesy of the City of Denver. Suppose this basic flow diagram represents approximately 1,000 different components of the system, including tanks, filters, piping, pumps, motors, sensors, and control systems.

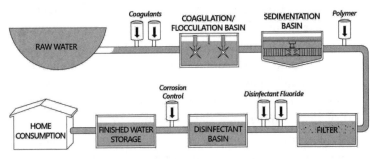

Figure 9 – Water Treatment Plant Overview

To keep the homes adequately supplied with water, the plant personnel will try to ensure that the various equipment items have effective strategies to maintain enough reliability. To do this, they will typically assess the probability and consequence of each piece of equipment, and ensure that the risk level of that asset never climbs above an acceptable level. Many organizations use a risk matrix to depict where the set of assets fails. The following risk matrix is a rudimentary example, with each dot representing a relative proximity to their risk limits. The dots represent individual assets in the system, and the shaded areas represent areas of unacceptable risk.

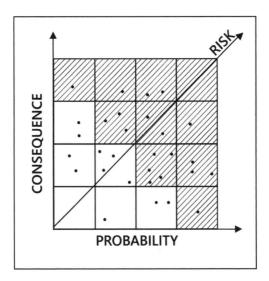

With each asset evaluated, the company will apply strategies to ensure that every asset is evaluated and maintained effectively within risk levels. However, this approach, looking at the components of a system, misses a key ingredient: it is the system we care about, not the components. Said differently, adding together all the individual risk levels of all the individual items could total substantially higher values than is reasonable. This happens because the probability and consequence of a single event on a single component is either limited or compounded by the connecting components, often in major ways. But looking at them individually misses this.

For example, assume we are evaluating a key one-thousand-feet-long section of pipe in the water plant. If you know the heaviest corrosion levels occurred in a specific, fifty-foot section, you would focus your attention exclusively on that section. Now, assume the corrosion rate on that pipe is such that the pipe wall could form a hole in as little as ten years, or as long as forty years. That means the uncertainty is spread over that time, and risk is climbing over that period. This could mean the plant personnel will begin working to reduce uncertainty, and therefore inspect, replace, or upgrade the pipe to reduce the uncertainty sometime ahead of that interval.

Now, instead of thinking of one fifty-foot section of that pipe, imagine this area of the water plant has a total of forty separate sections of pipe. If you look across the entire system, two *other* sections of pipe exhibit the highest corrosion rates and the highest probabilities of failure. When you look at the entire system, the risk of downtime is driven entirely by those two segments. If you know this, then you realize that doing anything to the first segment will not address risk at all. In fact, it will be wasting time and gathering data that does not bring additional precision to your decisions. You will only know that the risk levels of surrounding piping are negligible if you are looking at the entire system, instead of each asset independently.

To put it differently, looking at the entire system as a whole will easily identify those assets for which there is no *incremental* risk to the system. Comparatively, evaluating each asset individually will often produce an analysis that is substantially over-conservative, or one in which the key components of the system are underweighted. While this may not seem to be a problem, it is a major one. If results from risk analyses are overly conservative, it will appear that capital investments will pay off at a higher degree than they really will. When the recommended economic investment numbers are substantially higher than the total amount of profit for a facility for many years, no one believes in the analysis. The poor analysis will result in lower reliability. Only by looking at an entire system can we truly reliaball it.

As of 2020, almost no major process industries do this as standard practice.

This massive opportunity can only be realized if we drive substantial levels of quantification in both real data and algorithms, because this is the only way we can truly analyze complex systems.

Quantifying Complex Systems

Ten years ago, the idea of analyzing an entire baseball team with new data was a few years old, but still pushing the envelope. That is for a roster of twenty-five active players and fifteen more inactive. When they are on the field, a baseball team is indeed a complex system: every player is needed, doing different functions, at different times, in order to be successful. However, they were just using data to try to draft better players. Imagine trying to use computer algorithms for a team of a thousand players. Then imagine extending those algorithms beyond drafting players, instead trying to predict how they would perform.

In order to be considered a *complex system,* we need three things. First, the system must have at least 1,000 independent components/assets. Second, the system performance depends on all assets performing well. Third, the assets impact the performance of one another. In a large industrial facility like a water treatment facility or oil refinery, we may find 1,000 to 50,000 different pumps, pipes, and vessels that are necessary to make them run. Even more complex systems would be large corporations, where 100,000+ employees are the assets, or the Texas public education system, where 5.4 million students were enrolled in the 2017-18 school year.[1]

Looking back to our water treatment plant, if we assume 3,000 assets are running when the plant is in full operation, and we want to make determinations about optimizing the reliability of the system, how do we do it? Since we must focus on the overall system, we have to run and compile risk and performance numbers for all the assets together. Can you imagine asking a group of engineers and data analysts to run risk and performance calculations for 3,000 different assets? It would go something like this:

1. Compile the data for each asset and run a risk analysis.

2. Estimate a total risk level for the system.

3. Add the risk levels together for all the assets, to compare to the estimates for the system.

4. Review the results of each individual asset risk assessment against the others to determine the highest risk.

---

[1] "Enrollment in Texas Public Schools," Division of Research and Analysis Office of Academics Texas Education Agency, last updated August 2018, https://tea.texas.gov/sites/default/files/enroll_2017-18.pdf.

5. Weight adjust the assumptions and risk calculations for each asset to drive the cumulative risk to match that for the system.

6. Identify reliability improvements across the system, and repeat steps 1 through 5 as you go to determine the real impact of these improvements on the overall system.

If we had one hundred people working on it full time, it could possibly be done in a few weeks, but most facilities don't have anywhere near that amount of people or time to devote to such an endeavor. While this process might work in theory, practically speaking, this would never work, which is why it has not been done before.

Now, instead of 100 people, imagine that you had moved up the decision precision levels so that, for each asset, a quantitative risk/reliability algorithm was developed, and that each of these algorithms linked to one another. It wouldn't matter if you had 3,000, 30,000, or 300,000 assets. A computer could use the reliability of each asset to calculate the risk for the entire system in a matter of seconds.

"Easier said than done," one might say, and they would be right. That doesn't mean we shouldn't do it. While it will take a lot of work to focus our systems on refining our data, and our experts, scientists and engineers on building better algorithms, the massive opportunities here far outweigh the difficulty of the path.

The good news is, there is a path we can follow.

Chapter 9

How We Get There: A Different Roadmap

I F YOU WERE ASKED TO GUESS IN WHICH DECADES, since the Great Depression, the United States experienced the highest and lowest levels of productivity growth, how would you answer? If you remember history, you might assume the decade immediately after the Depression, which also included our involvement in WWII, would be the highest, as so many people went to work supporting the war effort. If so, you are right. The largest growth rate, measured in gross domestic product (GDP) per capita, was in the 1940s. Now, with the development of the internet, computers, and technology at our fingertips, you might guess that the last two decades would be up there too. You'd be wrong.

Over the last sixty years, GDP has grown relatively consistently. However, in order to understand true productivity, we must adjust for inflation. By taking GDP data from the World Bank[1] and adjusting for inflation levels based on the Consumer Price Index from the US Bureau of Labor Statistics,[2] we get the following numbers for total average economic growth by decade:

---

[1] "GGPD per capita (current US $) – United States," World Bank Group, https://data.worldbank.org/indicator/NY.GDP.PCAP.CD?end=2018&locations=US&start=1960).

[2] U. S. Bureau of Labor Statistics.

Table 5 – US Economic Growth By Decade

1960s	2.86%
1970s	1.44%
1980s	2.21%
1990s	1.67%
2000s	0.64%
2010s	0.86%

This means the average person in the United States increased their economic output by less than 1 percent per year for the last twenty years, and despite the unparalleled capabilities to learn and grow today, the last twenty years have seen the slowest growth in modern history. How do we explain this?

While there are many theories, I believe this: while technology has brought advancements in ways we never thought possible, it has also accelerated improvements in areas that affect our daily lives, creating staggering levels of comfort in modern society. The things from the 1980s and prior that humans could improve with a little ingenuity and some elbow grease have already been largely improved. The improvements to be made are no longer obvious, *and* we don't have much discomfort driving us to make changes.

Put another way, instead of technology pushing us forward, we have become content with the pace at which technology allows us to coast.

In that same vein, you may be thinking, "We've never had to do this *reliaball* thing before. Why now?" Over the past forty years, three things have changed. First, we have used our massive computing power to capitalize on most of the significant, acute opportunities we have faced—better cars, better energy efficiency, better technology. The challenges now are more complex, covering much larger portions of our economy, and requiring more sophisticated algorithms to solve. Second, we now have the ability to store and access terabytes of data easily, quickly, and affordably. While this would have cost hundreds of thousands of dollars just a few years ago, it is now pennies. Third, with computers doing *most* of the heavy lifting, we are left with the two things unique to humans: empathy and creativity (i.e., understanding a problem at its core [beyond correlation] and devising a solution outside of current resources).

While WWII was the spark to wake us up from the Great Depression, perhaps COVID-19 will be the catalyst to drive solutions we never thought possible. The biggest challenges our society will face, from unsustainable healthcare costs to more specific knowledge in our workforce, all depend on developing more sophisticated solutions than we've ever had. We need this complex reliaball thing. It is how we will make the biggest improvements in our society and will allow us to affect massive changes beyond what we've seen in recent history.

The Roadmap

How do we shift away from the old philosophy: *identify a problem, then get all the experts and data together in a room and solve it?* We've been doing this for a long time. It is going to be hard to convince people that this is less about *deciding* than it is about *managing data*, especially in large institutions like massive corporations and government entities. To be clear, we are not talking in this chapter about the quantitative decision process; we are talking about how to start applying it. In other words, what is the process of *shifting to* the quantitative decision process?

Luckily, while change is often hard, there are huge incentives (as described in Chapter 3), and the process is not as complex as one might think. If a group of people is willing to make the shift, when an opportunity arises, it comes down to ten steps.

1. **Identify a clear objective/problem.** As simple as it sounds, this is often missed. As discussed in Chapter 7, leaders must not allow a default like "be more reliable,"

"save lives," or "make money." We must be specific about what our objective is. "Minimize the overall *total* risk to people, including economic, health, and environmental."

2. **Establish criteria for success.** If one life was saved, were we successful? If the average equipment item ran 1 percent longer, were we successful? This must be identified. Statements like "the total fatality rate does not rise more than 0.01 percent over the next five years" or "boost the throughput of our facilities by 1 percent annually with a 30 percent rate of return," are specific and trackable.

3. **Identify required analyses.** In order to calculate risk, reliability, return on investment, etc., what calculations or evaluations will need to be done?

4. **Organize current available data.** Review all current data and determine where that data should be stored and managed so that it is most accessible for the analyses proposed.

5. **Validate experts.** We cannot simply accept a resume or a recommendation; we must ascertain the true level of expertise, and the amount of certainty we can place on their intuition.

6. **Calibrate uncertainties.** Where are there unknowns and gaps in abilities? How much will these impact the decision making? If one of these has a real possibility of skewing the results in a major way, you will have a large degree of uncertainty. If not, then uncertainty may be small.

7. **Perform initial risk calculations.** Given the current set of unknowns, data, experience/expertise, gaps, and uncertainty, how high is the risk associated with this opportunity or problem? At the end of this step, you need quantified risk numbers. Note: not having enough data or the right people does not mean you can't complete this; it means you must calibrate your uncertainty accordingly

in step 6.

8. **Empirical comparison.** Do the calculated risk levels compare reasonably with prior experience? For example, if you are calculating the risk levels associated with IT system outages at a company and the calculation came out to $50 million per year of risk, but the entire company's profit is $10 million per year, something is off. Even better, if the annual loss of productivity due to IT issues has been around 1 or 2 percent, which means between $1 and $2 million of downtime, there are big issues with the baseline. Either data, analysis, or uncertainty is skewed.

9. **Review.** Evaluate the process and identify any portion where things did not align.

10. **Plan.** As you go to the quantitative decision process, what needs to be done? Looking back at our diagram, which of these components needs to be bolstered to ensure that your data is being moved and processed in ways that ensure a solid outcome?

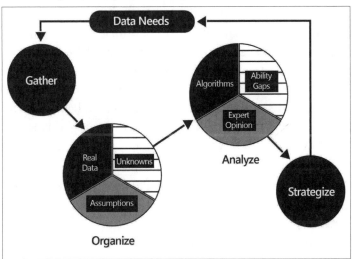

If your transition process has been effective, at the end, there should only be one primary question to be answered: What do we need to include in our plan to economically reduce uncertainty? The answers are in the diagram itself. We can do one or more of the following:

1. Go from gaps to better expertise to better algorithms.

2. Go from unknowns to assumptions to real data.

Gaps to Experts to Algorithms

If we simply don't have the right people or systems to do analysis, we have gaps. The size of those gaps depends on how fundamental that analysis is to our challenge or opportunity. For example, if you were looking to invest $100 million, typically you would want to put it into a number of different investments. When you consider the 3,600 stocks that are listed on US stock exchanges, their different industries, various economic indicators, bonds, treasuries, insurance, private equity, etc., there are literally millions of permutations. If you had never done this before, you would have a gap. You can close that gap by finding experts, using algorithms, or both.

A few years ago, I was sitting in a hearing in Austin, Texas, preparing to testify before the House Energy Committee. As I waited, I listened to several people come up to testify on the issue at hand, and as I heard their statements and read their bios, I realized that almost none of them really had expertise on the issue being discussed. Afterward, I was back in my office at the Railroad Commission, talking to a long time oil and gas operator. I lamented the fact that we didn't have the right people in the room that day, saying, "Why can't we get some actual experts to weigh in on this?"

"Do you know what an expert is, Ryan?" he asked. "It is anyone from more than fifty miles away, wearing a suit, and carrying a briefcase."

We both laughed, but unfortunately this humor was not as far from the truth as many of us would like to believe. When you ask yourself, "What makes someone an *expert?*" you may find yourself thinking: a lot of experience, seeing many different things, having substantial knowledge, maybe college degrees, possibly a long resume. Any number of markers may signify to us that they have a knowledge base or skill set to opine on an issue.

When I was in my thirties, I did expert witness work in court cases. With my experience in inspection, mechanical integrity, and risk analysis, I had a ton of expertise at a relatively young age. In fact, on one occasion, I was working with an attorney and they flat out told me, "Ryan, you have more knowledge on this than everyone else we've got; the only problem is you are too young. Juries don't believe you if you don't have gray hair." Fascinating.

This is not intended to belittle the time it takes to get experience. On the contrary, when that time is tied to repeated experiences with repeatable results, that is what translates into solid intuition. The challenge for us is to look past the simple measure of time, and ask how much of that time translated into relevant experience. Back to our investment scenario, you could find an economics professor at a university who had spent forty years studying the market, had written seven books on the subject, given hundreds of lectures, and been featured on business TV shows. A great resume should mean a great expert, right? What if he had never actually traded stocks? After him, you find a thirty-year-old who graduated from college, immediately went to work for a stockbroker, and in the past eight years has executed thousands of trades on hundreds of analyses, and had to manage the results. Which one sounds like more of an expert?

The trick in finding an expert is identifying what they are an expert IN. I recommend the criteria established by Klein and Kahneman from Chapter 5. In order for a person to develop a skill, the conditions met should be:

- An environment that is sufficiently regular to be predictable.
- An opportunity to learn these regularities through prolonged practice.

Extrapolating this out, we can create a simple test I will call the KahnKle (*konk'-le*) test.

- What is the specific area or event in which we are looking for insight?

- Has this person been in an environment where these events occurred with reasonably consistent results?

- Has this person been able to witness/experience enough of these event/result combinations that they can predict outcomes?

- What are the possible areas of weakness that undermine this person's expertise?

If we ran the KahnKle test on people we normally consider experts, we may find some surprising results. As we discussed in Chapter 5, Dr. Stephen Fauci, the most public medical expert during coronavirus, had focused his career on researching the immune system response to diseases. From the public records available, it appears he has had very little exposure to the spread and risk of a new virus. He would not pass the Kahnkle test. Similarly, if I ask a professor at my university about my ideas for a startup business, he may be willing to share a wide range of opinions on my business, but if he hasn't started and run a number of them, he doesn't pass the KahnKle test. Finally, if I am concerned about what might happen if a piece of pipe develops a leak in a refinery, I may connect with the company mechanical integrity expert who has witnessed and evaluated thousands of inspections and analyses on pipe condition. However, if they have only witnessed five failures, the KahnKle test tells us that he is not an expert in piping *failure*, he is an expert in piping *degradation*.

Using this test to ensure we know who our experts are, *and* what they are actually experts in, is vital. Once we know this, we identify the remaining gaps, and we can close those gaps with more experts, or algorithms.

Algorithms are combinations of rules, processes, and mathematical formulas that provide a singular output given a series of inputs. If I want to know the taxes that someone earning $1 million per year pays, I could ask an expert to estimate. A seasoned tax expert could probably quote a fairly specific range depending on some criteria such as deductions, family status, and passive income. However, I do not need an expert if I am willing to go to the tax code and work through the set of formulas and tables that determine the precise amount of tax that someone must pay. These tables and formulas are the algorithms.

Algorithms have been built for hundreds of years. Some are so common we forget that we once didn't have them. For example, we apply Isaac Newton's Second Law of Motion (Force = Mass x Acceleration) in all early physical science classes today, but when Newton published these laws in 1687, they were far from standard. Similarly, across the fields of mathematics and engineering, developments in the 1700s would become the basis for the Industrial Revolution, as algorithms enabled mankind to design and build things that were nearly impossible before, since only these algorithms allowed enough precision in calculations. This is the same issue we have today, with complex system problems outstripping the precision of our analyses.

The key to constructing solid algorithmic processes is to identify the inputs and outputs, then use any knowledge (empirical or theoretical) to calibrate the relationship. If input A goes up, and this drives output B to go up or down, why? More importantly, what is the relationship? If I pour a cup of water on the ground, it spreads out to a certain sized spot. If I pour 50 percent more water out, anyone can tell that the spot will be bigger, and could even estimate that it would be approximately 50 percent bigger in area. However, when it turns out that it is only 44 percent bigger, what happened? One could look at the parameters of the situation (absorption into the floor, friction of the flow, evaporation rates, surface tension, etc.) and attempt to calculate it using theoretical models. Or one could do enough experimental pours to establish a predictable relationship. Either way, once you know the formula that relates volume of water to size of spot, you have an algorithm.

In the world of quantitative decision making and decision precision, this is the ideal place to have experts spending their time. We often miss this because, in a single instance, it is much faster to get an expert to tell me what the required force is, or the tax rate is, or the size of water spot will be. However, if I have hundreds or thousands of these, and their functions are all interconnected, it is impossible to have the experts predict the results of the system. I need algorithms to calculate the complex situations.

Does that mean that algorithms are perfect? Of course not. Newton's laws of motion describe gravitational forces very accurately on earth, but become inaccurate when talking about large masses like planets and stars, which is when we must switch and apply Einstein's theories of relativity.[3] Regardless, a series of effective algorithms is still a far superior method to evaluate a complex system. Without them, each assumption, estimation, or opinion introduces another layer of uncertainty. Once several hundred components are combined, the uncertainty is too large to overcome, and the results are too risky. This is why our coronavirus reaction was overkill, and why refineries overcompensate for risk. Without good decision precision, and with an aversion to loss, leaders default to hyper-conservative decisions.

The right set of algorithms enables us to evaluate large systems efficiently, while managing uncertainty to a reasonable level. With them, our decisions and actions return the best performance. Since they are not easily developed, this is where experts can become exceptionally valuable, in both building the algorithms and providing the experiential/empirical check to validate the algorithm results.

---

[3] "When Did Isaac Newton Finally Fall," Forbes, Forbes Media LLC, updated on May 20, 2016, https://www.forbes.com/sites/startswithabang/2016/05/20/when-did-isaac-newton-finally-fail/#4a4f1a7848e7.

Unknowns to Assumptions to Real Data

Once we know what analyses we need to run, either through expert evaluation or algorithms, we know what inputs we need. While this sounds obvious, it is not the logic we normally use.

In most cases, in most industries, we have been gathering data for a long time. Doctors have been checking temperature, weight, heart rate, and blood pressure at every appointment for decades. However, when was the last time you saw a graph of these? For most of human history, the most relevant data was the data related to an object, event, or condition right in front of you. As a result, most of the data we gather is related to issues or needs that we have had in the past. If something was getting too thin, we measured thickness. If someone was getting too fat, we measured weight. If something gets too hot, we measure temperature. If someone works too long, we measure hours. This, of course, makes sense, and it has become the trigger for why we gather most of the data we do.

At the same time, is all that data equally useful? Of course not. In fact, is it possible that some of it is not useful at times? Perhaps never? We rarely ask these questions. Instead, it is easy to fall into the trap of *the data we gather is the data we use*. In fact, it is so easy to get comfortable with the current set of *analyses* on the current set of *data*, it is very difficult to move to a different method of evaluating either.

What this says about people is that we tend to treat the data we have as the available pool of data and we build our analyses around that pool. Here is a simple test for you. Read the nutritional information on the back of a package of food. What are you looking for? Calories? Fat grams? Carbs? Those are the three most common things people are looking for to gauge healthiness of the food. Take Nature Valley Granola Protein, Oats & Honey*, for instance. Sounds healthy, doesn't it? Here is the nutrition information from the Nature Valley website:[4]

---

[4] "Oats & Honey Granola," Nature Valley, General Mills, updated 2020, https://www.naturevalley.com/product/oats-n-honey-protein-granola/.

Serving Size	1/2 cup
Calories	210
Total Fat 4.5g	
Saturated Fat 0.5g	
Trans Fat 0g	
Polyunsaturated Fat 1g	
Monosaturated Fat 2.5	
Cholesterol 0g	
Sodium 135mg	
Potassium 125mg	
Carbohydrates 32g	
Dietary Fiber 3g	
Sugars 12g	
Protein 10g	
Daily Value of Protein	10%
Iron	10%
Calcium	2%

Is it healthy? A lot of people would say YES. 210 calories are not much. There are only 4.5 grams of fat. So, it's at least okay, right? I would say NO. Why? Because there are 12 grams of sugar. In general, people don't look at that line as much, and in fact, it is not, on its own, instructive. After all, an apple can have 15 to 20 grams of sugar, but I believe that added/refined sugar is notably less healthy than natural sugar, and the nutrition table doesn't say which this is. The ingredients list does: *Whole Grain Oats, Sugar, Soy, Protein Isolate, Honey, Rice Starch, Soy Lecithin, Salt, Baking Soda, Natural Flavor (mixed tocopherols) added to preserve freshness.*

The number two ingredient is sugar, so we can assume that virtually all that sugar is added, refined sugar, the unhealthiest thing we eat. No, I would say this product is not healthy. Is my analysis complete? No, but it is more thorough than what most people do before they eat something. It did require more than simply looking at the label. In fact, we could go further. What we would really like to know are things like glycemic index, effect on gut bacteria, or the level to which it causes insulin to spike. However, this data is not readily available, so we don't even think about using it to do a different analysis. This is the difference between real data and raw data. Obtaining and holding large amounts of data isn't helpful if it isn't the right data.

In this food example, we are driving at the core problem with data: we don't know what we don't know. Why don't we know this? Because there is plenty of data, but we don't know which of it is useful. Also, we have not stopped to consider what analyses we want to do based on the outcomes we are trying to achieve. If my goal is to eat a diet that maintains good physical health, boosts immunity, drops excess body fat, builds lean muscle, and drives good brain performance, I need to perform several analyses to build that diet. I need to analyze the right data, in the right form. This is *real* data. It is virtually impossible to perform these analyses based on the data provided in the "nutrition information" box on a food package. While I have plenty of *raw* data, I have more unknowns, and I am using more assumptions than *real* data. See below.

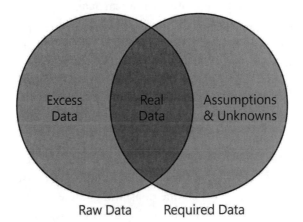

Unfortunately, we find ourselves in this situation more often than we care to admit. This is why most employee polls turn out to drive bad business decisions. When asking only a simplified set of questions, we get a set of simplified answers from employees. For example, a friend of mine discussed a poll with me that was done at his company, a mid-sized technology firm. In it, two questions were asked about productivity and culture. In both instances, 80 percent of the employees answered that working from home would help. The CEO took this and ran with it, and within four months, something like 60 percent of the employees were working from home on a regular basis. Two years later, productivity was down, and culture was suffering. Why?

At the two-year mark, the company brought in an organizational psychologist to work with their team, and she performed a number of detailed interviews on this specific question. After thirty interviews, she found an almost universal response. While the employees felt the company had good people, they also felt their teams were not well aligned. "Everyone here is just doing their own thing." The psychologist determined that people didn't really think *just* working from home would improve productivity or culture. They were reacting to poor teamwork with a desire to work from home. But, since the survey didn't ask this, their results didn't have that data. As a result, the CEO made a bad decision based on assumptions, not real data.

Understanding this disparity is a major part of removing uncertainty and increasing decision precision. As straightforward as this sounds, it is often a hard notion to employ. People who are data oriented will often want to start a decision or system evaluation process by gathering "baselines" or "initial data." This is not a bad practice, but it can lead to waste. When we launch into data gathering, we may spend time analyzing erroneous data or we may simply gather data that isn't useful. In a case where there appear to be urgent risks present in the short term and the data gathering process is time-consuming, there will be urgency to start. However, instead of starting by gathering a lot of data, it is better to start by determining what data is needed. From there, evaluating the data you have, identifying the unknowns and conservative assumptions, and quantifying the uncertainty in each will save time. We will certainly go find additional information to fill in known holes in our analyses, but it is rare that we think about analyses that are *possible* with data we don't even have. This is where the biggest opportunities can be missed.

In an ideal state, we are gathering the right data, and only the right data. This is not just about the numbers, but where they are taken, when they are taken, the quality of the readings, how they are calibrated, how they trend, and what sort of variation we see as they change. To make this shift, we identify the unknowns and assumptions that are driving the highest degrees of uncertainty, and use whatever tools are available to test and obtain more information. While this can seem daunting—shifting our data gathering practices, schedules, and techniques—it should almost always be a great front-end investment, as long as we can be certain that the real data meaningfully reduces uncertainty.

Calibrating Uncertainty

If you are thinking that this process, shifting from subjective to objective, could be time-consuming, you are correct. If you are thinking it may not be practical to gather enough real data or develop the complex algorithms to make 100 percent fully quantitative analyses, you are also correct. There is an important distinction between the data required to make *fully* quantitative decisions and the data required for *effectively* quantitative decisions. We want the latter, which is rarely the former. Therefore, understanding and calibrating the impact of what we don't know is critical.

Since our objective is to decrease risk associated with uncertainty, if increasing certainty does not decrease risk, then there is no value. How can this be? If we have effectively quantified and incorporated the amount of uncertainty (driven by unknowns, assumptions, gaps, or subjective expertise), we can calculate current risk *including* that uncertainty. In a large system, we commonly see the 90/10 rule at play: 90 percent of the risk is driven by 10 percent of the uncertainty. Therefore, if the risk is sufficiently low even *with* that uncertainty, there is no value in investing time or money to further quantify the process. Put another way, even though we may be able to increase precision around much of the system, it may not actually decrease risk. I cannot tell you how many times I have watched professionals who are looking for additional certainty go through major efforts to get data to try to overcome a problem, when the results offered no help at all. This is why calibrating risk and uncertainty across the system is so crucial to effectively quantifying the risks of the components.

One of the best, and certainly the simplest, ways to calibrate a model of the real world is to validate the results of that model against historical experiences. Believe it or not, this tends to be the exception, not the rule. At the pace of the modern-day world, leaders are looking to make decisions, complete the project, and execute the plan. After this is done, there is little value in going back and validating the tools, so why do it?

In THE FAILURE OF RISK MANAGEMENT, Hubbard makes very specific points about the lack of empirical validation in risk analysis. "Models that are built are rarely back tested," he wrote. "Comparing models to history is not itself a conclusive valida-tion, but models that do not fit historical reality at all are very likely flawed. If anyone had done this on certain financial mod-els, they would have known the 2008/9 financial crisis was much more likely than the models indicated."

This can go both ways. Calibrating a model or a calculation based on only one set of criteria can skew the model's outputs one way or another. For example, during the initial outbreak of coronavirus, multiple institutions were developing models to try to predict the spread of infection, the rate of hospitalization, and the mortality rate. The problem was that many of these models were trying to match their models to the results in Italy and New York City, the two worst-case scenarios. As a result, the models were skewed, predicting much worse scenarios than were actu-ally experienced. If these models had included results from South Korea or other parts of the United States, and made assumptions about social distancing in those areas, these conclusions may have led to less drastic, and less destructive, decisions. In addition, we could have allowed the disease to spread at a more measured pace, among low-risk individuals, then not shocked the system with a massive new wave when we reopened.

In a water plant, refinery, or mining operation, calibration is much easier. Calculate risk levels based on the current set of data, assumptions, expert opinion, and algorithms, then compare to history. For example, if someone calculates the risk across 10,000 assets in a water plant, and the average risk level for an asset is $50,000 of risk per year, the total risk across the facility is $500,000,000. If, over the past five years, the cumulative costs of all assets that actually did fail was $4,000,000, the original calculations were off by two orders of magnitude. If this was driven by uncertainty and bad assumptions, then instead of scrapping the analyses, either recalibrate the uncertainty or improve data or analysis to reduce the uncertainty.

How do we calculate/calibrate uncertainty when it comes to reliability and risk? The answer is *probability*. Given a certain set of inputs, one can calculate the probability of an event, such as:

- A 30 percent chance that a motor fails within the next three years
- A 1 percent chance of having a major car accident in the next five years
- A 25 percent chance of contracting coronavirus in the next twenty-four months
- A 4 percent, 18 percent, 54 percent, and 87 percent probability that a tank wall will develop a hole in the next ten, twenty, thirty, and forty years, respectively

In each of these cases, we are given probabilities based on a certain set of things we know and don't know. Often we begin with some nominal, or base rate. From there, the unknowns drive the probability up. If there was an average rate of 10,000 people a day contracting coronavirus in Texas, in a population of

10,000,000, It would take 1,000 days for everyone to catch the disease, and I would have a .1 percent chance of getting it on one day until that point, and a 36.5 percent chance that I would catch it in the first year. However, if somehow I knew I would contract coronavirus one year from today, what is the risk of contracting it in the next 364 days? Zero. Even though I know I'm going to get it sooner, the risk before that point is lower, because I know it. Certainty kills risk.

Unfortunately, we are never 100 percent certain, but we often have the ability to be more certain than we realize. Regardless, calibrating the uncertainty is done by establishing the range over which the uncertainty spreads events. For example, assume we took three measurements of a tank wall over twenty years at ten-year intervals, and the results were 0.50 inches thick at the start, 0.37 inches at year ten, and 0.30 inches at year twenty. The tank wall lost 0.13 inches in the first ten years, then 0.07 inches in the next ten. When will the tank wall fail? Somewhere between years eighteen and fifty-four. That span indicates the degree of uncertainty. If the probability of failure over that time adequately reflects the uncertainty, the uncertainty has been calibrated correctly.

This brings us to an important distinction. Lowering risk does not always mean increasing safety. If I jump out of a plane without a parachute, some would say that is extremely risky, because there is virtually 100 percent chance of death. But to be shrewd about it, there really isn't any risk. Up until the point I hit the ground, there is almost 0 percent chance of death; when I reach it there is 100 percent chance of my death. After I have reached it, there is no longer any risk; just the process of people around me reacting to my death. While this example is crude, it is important to differentiate between the event itself and the risk. We often place a disproportionately high-risk level on events that are more severe in nature. But increasing my certainty of a machine's degradation does not decrease its degradation, and simply knowing more about my health doesn't increase my chance of living. Only repairing the machine or making lifestyle changes will do that.

This also brings into focus another important distinction, the difference between random events and uncertainty. If I know that events are random, like weather, there is a fixed probability of them happening. Yes, there is still risk, and it is being tied to uncertainty, but the randomness means the event will eventually happen, and I must prepare for that eventuality appropriately, given whatever likelihood it has, or accept the results if they are infrequent enough. In Houston, every year presents a possibility of a major (Category 3, 4, or 5) hurricane. When you look at the number of Category 3, 4, or 5 hurricanes that have occurred, and the probability that they will hit Houston, you can expect a major hurricane event every four to five years. Therefore, homeowners, business leaders, and elected officials do not need to spend time trying to calculate the overall probability. Instead, their time should be spent determining the areas of highest exposure during those events (which could be treated as a system of components), weighing appropriate investment required to minimize the damage, and planning accordingly. In the case of a random probability of a component failure in a system, these probabilities must be included. Even though there may be no way to increase certainty with data or analysis, the analysis can still provide guidance as to how heavily to prepare for the eventuality.

Many risk analysis methods are based on a predetermined or normal level of uncertainty, which does not require the system user to calibrate. As the oil and gas industry rolled out Process Hazard Analysis (PHA), Failure Mode and Effect Analysis (FMEA), Reliability Centered Maintenance (RCM), Risk-Based Inspection (RBI), and other methods for assessing and managing risk, uncertainty was, at best, a consideration. Moving to the future of quantitative system analysis, uncertainty and decision precision must transition from a casual piece of our calculations, to the focus. This will be more than just a tool or a process change:

it will be a change in the way our people think about the process of system design and assessment. It will be a change in culture.

Establish a Culture

But I am not a data scientist, engineer, or economist. I don't have a background in statistics or software development. What can I do to quantify decisions if I don't know how to gather, organize or analyze data, or create a strategy to use the results? It's a good question, and one that anyone who is not in the *data* space is asking themselves right now, whether they know it or not. How do I know? Fear. As we discussed in Chapter 5, at the base of our motivation is fear, and once triggered, it trumps everything else.

Over the past 150 years, there has been a continual cycle of workers being employed to do a job, and once that job gets more repeatable, a machine is built to do the task. Then workers who previously did that job must find a new place to apply their skills or learn new ones. When we talk about automation, machine learning, and artificial intelligence, some people get excited. They see the opportunity to advance. Others get scared. They hear estimations that as many as thirty-six million jobs could be replaced by automation in the next two years,[5] and fear becoming obsolete. They are asking themselves, "How do I participate in this?!?"

The answer is *leadership*. While repeatable tasks may be easily replaced by machines, creativity and empathy cannot be. Machines can learn, but they don't learn like humans. Our ability to take in information, grow in our abilities, and apply new knowledge and skills in creative ways remains a uniquely human contribution. This will likely always be where the big, bold ideas and new solutions come from. Software and algorithms may recognize patterns and find the best option out of thousands, but only humans will arrive at the crazy new directions we've never

---

[5] "Over 30 million U.S. workers will lose their jobs because of AI," MarketWatch, MarketWatch Inc, January 24, 2019, https://www.marketwatch.com/story/ai-is-set-to-replace-36-million-us-workers-2019-01-24.

tried before. This is how humans beat machines. However, we can lose if we fall out of the cycle of growth, and begin to prioritize status (social rank, titles, money, and possessions) over impact. Some of the concepts we have discussed here, like challenging current medical practices, applying data to personnel performance evaluation, and asking experts to do part of their job differently, will be difficult changes for people. Therefore, it is incumbent upon leaders to establish a culture of growth, and in particular, growth in how we use data in our lives.

How do we do this? By establishing an expectation that we continuously bring data into the discussion, and do not accept the prior historical consensus without it. In the next chapter, we will walk through some examples to help explain, but simply put, every decision requires data. If your child, or your spouse, or your coworker, or your employee brings a need to you, ask them for data. When they don't have it, ask what data they'd like to have to make the decision. Ask how they might weigh decisions based on that data. Most importantly (and this is also the hardest), when there is one set of data that people have applied for a while to make decisions, ask them to take a step back and apply the ten steps at the beginning of this chapter. Ask questions.

1. What is our objective, and what does success look like (specifically)?
2. What analyses could inform us as to the best path?
3. What data do we need to do those analyses, and can we get it?
4. What is missing (uncertain) in what we are talking about?
5. What are our choices, and how can you identify which is the best?

These questions feed into the ten steps, and asking them repeatedly will get others in the habit of answering them before you ask. It will also get them to start asking them as well. In addition, for a culture to be focused on anything, including data analytics, it is important to ensure *philosophical* alignment. This can be done by developing some rules, or statements, that guide how your team approaches these kinds of situations. If done correctly, these would be considered any time someone is evaluating a situation and applying their resources. Here are some examples:

- Getting data is not the objective; reducing uncertainty is.
- Data is never wrong; it is incomplete.
- Intuition and data, if applied correctly, cannot disagree. If there is misalignment, we need more of one or both.
- We don't want people to make decisions; we want them to fill in the uncertainty in our decision process.
- We don't need to know we have the best experts; we need to know what they are experts in.
- It is okay to make a judgment call, as long as we have calibrated the uncertainty, and therefore the risk, in that judgment.

These are not intended to be hard and fast laws. Instead, they should accentuate key thought processes that differentiate a new mindset from a traditional one. Since organizations will be in different phases of applying quantitative decision making, it is important to develop a set of principles that help move their group forward.

If a group of people is accustomed to their supervisor, coach, or parent always starting with these principles and asking the same questions, they will adapt themselves to this way of thinking. They will begin to align with those principles and ask those questions before their supervisor does. The larger the organization, the bolder leaders must be to make a shift. But as we have seen, these are also the groups with the greatest potential to gain.

Since this thought process is new for most people, this culture shift will be a big step in making the world reliaball.

Chapter 10

A Different World:
Reliaball Examples

IN 2007, I WALKED INTO A LARGE CONFERENCE ROOM at a North American refinery. The VP responsible for that plant and a number of others were in the room, as well as the plant manager and the plant leadership team. The issue on the table was whether to postpone a turnaround (shut down for maintenance work) that was scheduled to start in a few weeks. Two days before this meeting, a key compressor in one of the units had failed, and the unit was taken down. This brought several other units to reduced rates. The maintenance personnel were scrambling to repair the compressor, and there was an opportunity to do a few other key things at this point that would have been done during the turnaround to keep the plant running longer. At the same time all this was happening, refining margins for that plant had reached some of the highest levels they had ever seen.

There were around thirty people in the room, including people from turnaround planning, marketing, operations, maintenance, inspection, and engineering. There were also several experts in the room—machinery, process, corrosion, and safety. I was thirty-one years old at the time, and my eight years of experience was not vital to the discussion. However, I had been setting up a mechanical integrity system for the company and was in the room to advise on the information in that system. While we didn't use this term at the time, I was the data analyst. If you have seen the movie *Moneyball*, I was the nervous young kid in the room listening to the Scouts cuss and discuss. Listening, that is, until the plant manager turned to me and said, "You're the inspection engineer, right? What do you think?"

I told the group what I was seeing in the data, and that the risks presented by delaying the turnaround were very easily managed. Unlike *Moneyball*, in this case I was not at odds with everyone else in the room. But they were at odds with each other. Around the table, I watched as various groups and experts argued their point and explained their view of the risks. They all brought some information, mostly anecdotal, to explain their position. From potential cost impacts to equipment failure issues to "we don't know what we don't know," everyone had an opinion. A couple of people even got emotional arguing with each other about how they needed to "stick with the plan" or how this wasn't the right time to "prioritize business over safety." I watched as the plant manager and the VP listened intently as everyone spoke. They asked very specific questions, took notes, and challenged. By the end of a two-hour discussion, they made the *crucial decision* to delay the turnaround for nine months.

A year later, this turned out to be a great call. The plant ran relatively smoothly for the next nine months, and by the time they shut the plant down for the turnaround in 2008, their margins were much lower. They made hay while the sun shined. But what if margins had improved over that time? What if the plant had experienced other problems instead of running smoothly? So many things were based on a range of opinions and pieces of information. Even at that early stage of my career, I could see there was uncertainty in the decision, and that the uncertainty was the real risk. In fact, I remember thinking if that crucial decision had been mine to make, I might have folded under the pressure.

Imagine that you were sitting in a room like this. Instead of a refinery, maybe you are facing a crucial decision to hire fifty new employees. Maybe you are trying to figure out how to run a business in a downturn. Maybe you are dealing with complex health issues with a family member. Maybe you are trying to change education policy to prepare our nation's workforce for the next ten years. Or maybe you are the President of the United States, and there is a new virus spreading in your population. Instead of facing these challenges with a group of experts offering opinions, what if it looked like this? You walk into that conference room, and there are thirty people in the room. The problem, objective, and success statement are written on the wall. Someone in the room is showing a model that integrates all the complex variables, decision possibilities, algorithms, and data. Having already identified the values or points at which different actions would be triggered, you are now working as a group, running the calculations to determine what the resulting action should be.

As you work around the room, people are adding their input to address the specific areas in which they have experience. Data is being captured and added into the model, and the outputs are changing real-time as this happens. The screen in front of everyone is not only showing risks or highlights but identifying areas of uncertainty. The group is working together to tighten up the uncertainty, trying to find or capture more data, and calling for input from people that aren't in the room to provide even more granular insight. In the end, the model gives you an answer. A number. Because the number is lower than your threshold, you not only make the crucial decision, but you know the probability of outcomes as a result of that decision.

For example, you hire this person and make them the project manager for this new project, knowing they have an 81 percent probability of being successful. Or, you conclude that your aging parent has a fungal overgrowth in their gut, and that diet changes and certain natural supplements have a 94 percent chance of helping them lose weight, sleep better, and have more energy. You determine that postponing the turnaround is the right move, with a 74 percent chance of coming out better after a year. Or, you decide that quarantining a small subset of the population with preexisting conditions has the overall lowest negative impact on Americans.

The analysis indicated the range of actions that have the highest probability of success, and the degree to which success was likely. That is what the future will hold. We are already seeing some striking examples of this playing out in the world around us. The challenge is that today most people think of data analysis as a specific set of tasks assigned to people with specific titles like "data scientist." This is one of many things that must change. Any leader faced with a crucial decision in a complex situation can demand that her team apply more logic and data as opposed to pushing for them to simply "make a call." Making the call is sometimes faster, but dealing with the wrong decision is not. To get you thinking, let's walk through some real examples of crucial decisions in complex situations and systems, and how they were improved with a different analysis of data.

Case Study on Hiring and Social Media

At the beginning of 2019, a company in the Houston area was looking to expand its hiring, targeting the addition of nearly two hundred people. Leading into this year, the company had experienced consistent growth, but also high turnover. The company had a team of in-house recruiters who had many years of combined experience. They worked tirelessly to identify and bring in the right people, and in general, they were hitting their numbers.

A few weeks into the process, a few candidates began to drop out of the recruiting process. They cited reasons like: they were going to try and make it work where they were, or salary was not high enough, or more general things like "personal situations." This uptick in people dropping out of the process was abnormal, so the recruiters were looking for answers. They checked in with a couple of outside recruiting firms they worked with on occasion, and both firms mentioned that the company's *Glassdoor* rating was "really bad."

Glassdoor.com is a social media site designed to allow employees of companies to leave anonymous reviews on the companies, including personal opinions, experiences, etc. The site started up in 2008, and by 2017, had 41 million users. In 2018, the site was acquired by a Japanese company for $2.1 billion. Their primary revenue source is from companies that pay the site to advertise for potential hires or build a company profile that allows them to respond to negative reviews.

The company in our example had 155 reviews, and an average rating of 2.6 out of 5.0 stars, a low rating, at the time of this study. The three most recent reviews were extremely negative, saying things that were not only harsh, but in two of the three, saying things that were simply untrue about the company. The recruiting team became very concerned that potential employees would read these statements and turn away from the hiring process. As a result, they began to discuss what needed to be done to solve the "Glassdoor problem." There was talk of advertising, paying for a Glassdoor membership to address the comments, hiring a specific firm to help address this problem, and asking current employees to do positive reviews to counteract the negative reviews. The total cost was stretching into tens of thousands of dollars, not to mention the distraction of driving current employees to give positive reviews.

A crucial decision fell to the company leadership. *Do they attack this Glassdoor problem in this way? Do they do nothing and allow Glassdoor to continue to have its negative impact? Or do they try other routes to get a better result?*

Outside the recruiting team, one of the company's data analysts heard about the quandary and took it upon himself to do an analysis. First, he defined not one, but two objectives.

Objective 1. Determine if there is anything we can learn from Glassdoor about how we can improve our company (i.e., reviews from employees with good ideas we need to implement).

Objective 2. Determine the level of negative impact that Glassdoor is having on the company and what would be needed to minimize that impact.

To evaluate, he looked at the last year of reviews, forty-two total. On the surface, the data confirmed the early conclusions that Glassdoor indicated problems. These reviews had an average rating of 2.1 (low). However, since there were nearly a thousand employees at the company, he recognized the limitations of this data set, and went further.

Delving into each individual review, he extracted the length of the review, the specific compliments or complaints, the job title of the person leaving the review, and the employment (current or past) of the person. It should be noted that all Glassdoor reviews are anonymous and unverified, so there is no way to know if information presented is factual or fabricated. Reviewing the data on face value, he found the following:

Employment Status	Number of Reviews	Average Rating
Current Employee	8 to 10	3.6
Left Voluntarily	7 to 11	3.0
Terminated	19 to 25	1.2
Never Worked	2	2.0

The analyst developed and organized a much more complete set of real data and performed a fairly simple analysis. He concluded that there was such a small number of reviews from current employees (less than 1 percent of the total employee count) that these reviews were not a good representative sample of the entire company. Since most reviews were left by former employees, there was little value in reacting to their feedback. Therefore, the conclusion for objective #1 was NO. This was not valuable input to use in shaping the future of the company.

Objective #2 was more difficult. While the company was generally hitting its recruiting targets, it was possible that the process was being made more difficult by Glassdoor. But how could he tell if these negative reviews were having an impact on the recruiting process? To assess this, the analyst had to pull in additional data, and he landed on one more number that most people looking at Glassdoor would ignore. Any time someone leaves a review on Glassdoor, others who read the review can click a "helpful" button, and the number of times this has been clicked for a review shows up on the rating. The analyst concluded that he could identify the impact of reviews by referencing this data. The following is what he found when comparing the rating level and helpful clicks:

Star Rating	Avg. Number of Helpful Clicks
5	0.7
4	0.8
3	3.3
2	16.4
1	14.2

To explain, the reviews with five stars (positive) were clicked as "helpful" less than one time per review. The reviews that were one or two stars (negative) were clicked as helpful an average of more than fifteen times. Anecdotally, he noticed that one three-star review that was extremely detailed and had some very good specifics, positive and negative, had a total of five helpful clicks. A one-star review with a total of twenty-four words simply said the company was awful, the leaders were fools, and the company would soon be bankrupt. It had twenty-two helpful clicks. When it came to objective two, the analyst concluded that this Glassdoor site had become a site for former, disgruntled employees to vent.

When the issue first came up, the desire from some of the recruiting and leadership staff was to commit to doing the things needed to get the Glassdoor reviews up. A crucial decision was put in front of company leadership: to invest tens of thousands of dollars and create a major distraction or accept a more difficult and expensive recruiting process. However, these choices were based on a limited set of data. The data "experts" were only looking at the negative reviews (that Glassdoor wanted companies to see). After applying some decision precision, getting better data, and performing better analysis, there was almost no decision to make. The conclusion was to do nothing, as spending effort and money addressing the Glassdoor problem would have no meaningful positive impact.

This is an example of how crucial decisions can work in a very specific instance. But what about large scale decisions on more macro level challenges?

Case Study on Philanthropic Donations

What could be broader than the way our society donates, allocates, and utilizes philanthropic donations by people and corporations? It literally spans the entire globe and touches nearly every level of societies. But, is it doing any good?

This question relates back to one of the fundamental questions we struggle with in quantifying risk, reliability, and impact. What is the value of human life? But instead of thinking about it as an absolute measure (whether to spend money), think of it as a prioritization question (how to spend the money we have). As we explored in Chapter 1, the United States burned through $5 trillion in March, April, and May in reaction to coronavirus. When we weigh that amount against the normal value assigned to years of quality life, this was a minimum of ten times more than the "normal" value, but probably closer to a hundred times more.

Some people say, "Oh, you can't put a value on human life," thereby implying that all lives have the same value. I find this a ludicrous statement. While the idea of placing a value on human life may sound like a callous notion, let me make it more personal. At the time of writing this book, I am forty-five years old. I am in a great state of health, authoring books, running companies, and serving in public office. However, I would say that my life is not as valuable as my children's and would gladly give my life to save theirs if the choice was required. In the same notion, I would say that my mother, a brilliant woman who had five degrees, taught chemistry, and provided spiritual counseling to hundreds for free, has lived most of her life with great impact. However, she is now seventy-six years old and she is suffering from Alzheimer's. During COVID-19, our family had to deal with the emotional weight of watching her struggle in her final phase of life, isolated in a facility in which we couldn't visit her during the lockdowns. While I am incredibly saddened by the knowledge of her coming passing, her remaining life is not worth near what mine is.

Going back to philanthropy, we have limited resources, and we want to have the biggest possible impact on the people of the world with those resources. So the question we must ask is, "What can we do with that money?" For example, if there was a way to invest $50 billion (1 percent of the coronavirus losses) to save or improve the lives of 1 million people, wouldn't that be better than $5 trillion to allow 250,000 to live an average of an extra couple of years? I know this seems callous or an almost impossible decision to make, but they are the real decisions that people are making every day. The fact is, without data or a quantified analysis, people will tend to do what they have normally done, or what is comfortable, and often never realize what else might have been. Philanthropy might be the clearest example of this on earth.

In 2017, over $400 billion was donated to charities,[1] most of which are designed to either save or improve lives. Some of the US charities with the highest receipts, like the United Way, Task Force for Global Health, Feeding America, Salvation Army, and St. Jude's Children's Research Center (the top five charities donated to in the United States) focus heavily on the impact on human life. Between the five, they took in over $12 billion,[2] and a quick review

---

[1] "Giving Statistics," Charity Navigator, John. P. Dugan, updated 2018, https://www.charitynavigator.org/index.cfm?bay=content.view&cpid=42.

[2] "The Largest U.S. Charities For 2017," Forbes, Forbes Media LLC, updated on December 13, 2017, https://www.forbes.com/sites/williampbarrett/2017/12/13/the-largest-u-s-charities-for-2017/#62ea1f0f58e1.

of their public reports indicates they have helped, saved, or otherwise served over 5 million people. How many of these lives were "saved" is a nearly impossible number to determine. However, if we limit our number to those who received better nutrition and healthcare, and therefore experienced a notable increase in long-term quality of life, the number is at least 1 million.

The problem is, we often don't know how effective our charitable giving is. In a Forbes article from 2012 entitled, "Why Your Charitable Donations Probably Aren't Doing Much Good," the author explains that most charities don't end up delivering results because donors "tend to give to organizations that have brand recognition or simply because they request money from us . . . all without asking whether our donation will accomplish its goals." One example is particularly pointed.

> It takes $42,000 to train a guide dog to help a blind person, according to Guide Dogs of America. But to really help the blind, you could put that $42,000 toward funding eye surgeries for people in Africa suffering from a bacterial eye infection called trachoma. Since surgery costs as little as $25 and is 80% effective, you could theoretically restore the sight of 1,344 people with that $42,000. As The New York Times puts it in an article on effective giving: "If you value all lives equally—and in a minute I'll get to the fact that we certainly don't—then if you are training a guide dog, you might as well be giving to a charity that wastes 99.93 percent of its money. (Actually even more, as a guide dog does not restore sight.)"[3]

In WHEN HELPING HURTS, primary author Brian Fikkert explores this topic in detail. He theorizes that most charities are ineffective in actually achieving positive results because they focus largely on the amount of money that is exchanged or on a specific function that is highly subjective in its value, like

---

[3] "Why Your Charitable Donations Probably Aren't Doing Much Good," Forbes, Forbes LLC, December 14, 2012, https://www.forbes.com/sites/learnvest/2012/12/14/why-your-charitable-donations-probably-arent-doing-much-good/#4ad68bb74278.

education. Instead, he says, we need to focus on the core areas of happiness and fulfillment, the absence of which he calls *poverty of being*. When it comes to charities, he writes, "One of the biggest problems in many poverty-alleviation efforts is that their design and implementation exacerbates the poverty of being of the economically rich, their god-complexes, and the poverty of being of the economically poor, their feelings of inferiority and shame."[4]

Why is the world wasting so much money on charities that may be doing more harm than good? In a word, data. We tend to donate to a cause that has a good story or a good message, or because we have a personal connection to it. As a result, we may not approach the decision as a crucial one. As such, we either accept that there is not good data to measure outcomes, or we use the small amount of available data as a measure. For example, one piece of data I am guilty of using is whether the charity has low overhead. Certainly, low overhead tells me that a charity isn't wasting money on fancy offices, but that really tells me nothing about whether they are actually being effective.

Some people will tend to accept the classic notion that we can't measure the positive impacts that getting involved has on society. The Center for High Impact Philanthropy at the University of Pennsylvania disagrees. On their website, they say, "When people say an impact can't be measured, they often mean that they can't practically measure that impact to the degree of accuracy they want and within the timeframe and budget they have. However, saying something is impossible to measure makes it dangerously easy for donors and nonprofits to be let off the hook. What's more, it can lead to a focus on what is easy to measure as opposed to what matters."[5]

---

[4] Steve Corbett, and Brian Fikkert, WHEN HELPING HURTS: HOW TO ALLEVIATE POVERTY WITHOUT HURTING THE POOR... AND YOURSELF (Chicago: Moody Publishers), 62.

[5] "Impact Myths: Some Impacts Simply Can't Be Measured," The Center for High Impact Philanthropy, University of Pennsylvania, updated November 1, 2103, https://www.impact.upenn.edu/impact_myths_some_impacts_simply_cant_be_measured.

In other words, it is easier to go with our gut, since this decision is not that crucial to our own lives. However, if we were on the receiving end of the tens of billions of dollars in charitable giving every year, you can bet we would view these decisions as very crucial. As more companies and individuals are trying to focus their philanthropic efforts in places they believe truly make an impact, they are looking more and more for a solution to this problem.

To attack this, organizations began to spring up in the late 2000s that assessed the impact of charitable organizations. Give-Well started in 2007 and has since performed evaluations of hundreds of charities around the world. Their focus is to identify the charities that have the largest impact on the greatest number of people proportional to the amount of money they spend. Per their website, "We think that many of the problems charities aim to address are extremely difficult problems that foundations, governments, and experts have struggled with for decades. Many well-funded, well-executed, logical programs haven't had the desired results."[6]

But they have added decision precision to this area by looking at important data and building a set of algorithms to evaluate this data. Not satisfied with the old measure of low overhead, they have established new measures in four key areas: *evidence of effectiveness, cost-effectiveness, room for more funding,* and *transparency.* While their assessments do include subjective analysis by their team, they have developed true expertise by evaluating so many charities over the years. And, they apply both quantitative measures and comparative analysis between charities to eliminate uncertainty in their approach. How do we know? Their website has a section devoted to "our mistakes" which explains lessons learned over the years.

The result of these quantitative measures is astounding. The Forbes article mentioned here referenced the work of Toby Ord, a

---

[6] "The Wrong Donation Can Accomplish Nothing," Give Well, The Clear Fund, https://www.givewell.org/giving101/The-Wrong-Donation-Can-Accomplish-Nothing.

researcher in moral philosophy at Oxford University. He wanted to find out how much charitable projects differed in terms of their impact in an area such as health. By working with organizations like GiveWell, he found that some projects had an impact one hundred times (10,000 percent) bigger than others.

When you begin to think about $400 billion invested across tens of millions of lives, the opportunity is mind-boggling. But so is the complexity. Think about these questions: *How are the needs in Bolivia unique and different than the needs in Malawi? What are the positive and negative impacts of providing clothing and shelter versus a better education? If someone does get an education, are there any jobs or economic opportunities afterward? If we drill water wells in a third world country, do the people know how to maintain them? If a society has been decimated by AIDS, drought, or genocide, are there things that must happen before aid even begins? What things provide short term relief, while others have a lasting impact? More generally, where should money be invested? How should it be invested? Who should be prioritized? What research, infrastructure, or staff is needed to make the investment effective?* All these questions speak to a truly complex set of systems, with a set of decisions currently being made with a generally low application of data, analysis, and quantitative strategy.

If we consider the size and scope of all philanthropic efforts, as well as the potential positive and negative impacts if done well or poorly, the decisions couldn't be more crucial. Through more real data, better organization, better analysis, and better strategy, it is not hard to imagine a time when a donor can easily tell which efforts have the greatest positive impact. Imagine how much impact $400 billion could have if it was all focused in the most effective areas.

This philanthropic case provides an example of one of the most far-reaching complex situations, but we don't have to look much further than our own houses to see how data can affect decision precision and reliability.

Case Study on Air Conditioning Systems

Approximately 102.8 million homes had air conditioning in the United States in 2015.[7] That same year, there were 4.5 million new AC systems shipped in the US.[8] Of those, 500,000 went to new homes that were built that year.[9] This means that 4 percent of homes replaced their entire AC system. In addition, another 10 million homes replaced a portion of the AC system.[10] In short, 14 percent of US homeowners replace a portion of the AC system every year, to the tune of several thousand dollars.

Imagine. It's Saturday afternoon in the middle of June. You notice the temperature in your house is climbing into the high seventies in the middle of the day, despite the thermostat being set to seventy-three. Last summer, this wasn't a problem, so something must be wrong. You call the AC guy. Fortunately, you have a maintenance agreement with this company for them to come out and perform semi-annual maintenance. They move you up on the list and put you on the schedule for Wednesday afternoon. The technician comes out and finds that your unit has a valve freezing up due to a Freon leak. He can order the parts and do the repair, but they are booked for the next two weeks. He can classify it as urgent, and come back out over the following weekend, but it would be considered a rush job, i.e., more expensive. The parts and labor for the technician to come back out and perform the replacement job on Saturday is $1,500. If they can wait and do it on a normal schedule, it will be $800.

[7] "Table HC7.9 Air conditioning in U.S. homes by home size, 2015," U.S. Energy Information Administration, updated May 2018, https://www.eia.gov/consumption/residential/data/2015/hc/php/hc7.9.php.

[8] "Air conditioners shipments in the U.S. from 2001 to 2019," Statista, Dominican University, updated March 2, 2020, https://www.statista.com/statistics/220357/manufactured-shipments-of-unitary-air-conditioners/.

[9] "Number of new houses sold in the United States from 1995 to 2019," Statista, Dominican University, updated May 20, 2020, https://www.statista.com/statistics/219963/number-of-us-house-sales/.

[10] "U.S. HVAC Systems Market Size, Share & Trends Analysis Report By Product (AC, Heat Pump), By Technology (Inverter, Non-Inverter), By End Use (Industrial, Commercial, Residential), And Segment Forecasts, 2019-2025," Grand View Research, Grand View research, Inc., November, 2019, https://www.grandviewresearch.com/industry-analysis/us-hvac-systems-market.

You now have a crucial decision. While it may seem obvious—we must have AC in the summertime—for many people, this is a big problem. Do you spend the money to do the job on a rush, or take a chance that it could fail completely in a couple of days and you go a couple of weeks without AC? To make matters worse, if you had caught it when the Freon was beginning to get low, it would have just been a recharge of $200, but now it's $600 to $1300 more for the extra parts and the labor.

This scenario is playing out all over the US every summer. Of the 10 million homes that are replacing a component of their system every year, it is estimated that half to two-thirds of those are due to problems only being detected when they affect the entire system. For most people, their home is the most valuable/expensive thing they own. But if you consider a home as a collection of different pieces, the AC system is the most expensive to maintain. In fact, when you consider depreciation and maintenance costs, for most Americans, air conditioners are the second most expensive asset they own next to their cars. Even with that, most of us maintain our AC systems like they were refineries forty years ago. The only data we have is the temperature in our house, which simply says whether the equipment is running. As a result, we only get to make the crucial decisions when it is too late.

In 2019, a company was started in the Houston area called SmartAC. Their technology expands the amount of data used to track home AC systems, to give the owner much more insight into the performance of their machines. The model is incredibly affordable. For $99 and $5 per month, they can outfit a home with the sensors that track pressure differential across filters, temperature differential across the unit, and condensation level in the drain pan. Using their algorithms, they use this data to predict when units are beginning to develop problems long before they become major issues. This has the potential to save hundreds or even thousands of dollars by preventing major equipment issues or performing repairs on a planned versus emergency basis.

Current estimates are that the US on average is spending close to $15 billion on AC systems every year, plus an additional cost of $5 to $10 billion in services. If better data and better analysis can extend the average life of a home air conditioning system by 10 percent, the opportunities for savings are impressive. While a $100 to $200 reduction in average home AC costs may not seem large, $10 billion in reductions across the entire US population is staggering.

I use this example because it is close to home. Every one of us has burned with frustration (pun intended) at the thought of making an emergency AC repair. We never feel like we have good information (not enough data), and we are forced to make a judgment call based on a bit of consultation with one expert. We don't have access to software programs and data scientists, but we can choose to increase our decision precision. In the end, AC system choices rarely have life or death, or even long-term implications, but SmartAC will still be bringing a whole new level of data gathering, organization, analysis, and strategy to an individual's crucial decisions, and they will reliaball our home AC systems in the process.

What it All Means

The world is amassing more and more data all the time. Some estimates are that the world will produce 50 zettabytes of data (50 trillion gigabytes) in 2020, which will be over one-third of all the data generated in all of history before 2020.[11] Yet, as we explored in Chapter 9, with all that data, our improvement (measured in economic dynamism) is slowing. Why? It's not that we don't have enough data; we probably have thousands or millions of times as much data as we need. The problem is, with all this data, we don't know what to do with it.

As we move into the next few years, our society is facing challenges of a complexity it has never before faced. How do we make

---

[11] "Volume of data/information created worldwide form 2010 to 2024," Statista, Dominican University, updated May 2020, https://www.statista.com/statistics/871513/worldwide-data-created/.

enough food to feed a global population growing exponentially? How massive will the impacts be from global warming (whether natural or anthropogenic), and how will we manage them? How can we continue a healthcare system that is skyrocketing in costs, but we don't have the resources to pay for it? How do we evolve our education system to move us into the next decade, as machines, automation, and systems change the job landscape dramatically? Each of these challenges, and dozens more like it, present the same big obstacle. These are highly complex systems with hundreds of thousands of variables and inputs, and millions of permutations. The only way to develop the optimal plans to solve them is by using more sophisticated analyses than we have ever used before. There are no better examples than those in government.

When the Declaration of Independence was signed, the United States had a total population of around 2 million people. That means that every congressman (around sixty) represented roughly 33,000 people. At that time, most Americans were working as farmers or skilled tradesmen. There was no public healthcare, public education, social security, or welfare. In short, while they were doing important things, those things were simple in nature. Today, with 330 million people, 435 congressmen each represent nearly 800,000 people. Different states and regions have vastly different industries, economies, and social structures. Federal, state, and local governments combine to form one-third of the US economy, and they manage everything from the military to education, from healthcare to road construction, and from NASA to social security programs. In other words, over the last 250 years, while the population has grown by a factor of 100, our government has grown by a factor of 100,000 in both size and complexity.

Yet by applying the methods of old, our government struggles to be effective. Instead of developing the best solutions, the answer is almost always "more." In reaction to COVID-19, the PPP stimulus package injected hundreds of billions of dollars of aid. If you were lucky to get the aid money, you could keep your employees on staff. But if you weren't, you were at a massive disadvantage. Coming out of the lockdowns, companies ranging from restaurants to energy companies to athletics facilities are going bankrupt while their competitors will survive, based largely on these loans.

The conclusion? Rash moves by government will always have a destabilizing effect.

Instead of these types of reactions, when it comes to national disasters or acts of God, as they are called, our governments need to move out of the support business, and into the insurance business. While we can certainly mobilize personnel when needed, the funding of business and institutions after a hurricane, tornado, flood, freeze, or pandemic should be part of a modern rainy-day fund. With more complex modeling and sophisticated data analytics, we can implement the structures to capture, save, and disperse those funds quickly and easily when needed. This would not only ensure that government involvement is smooth and efficient but would prevent a rapid response from destabilizing the business environment.

In education, the problem is more acute. 76 million[12] kids go to school every year in the US, with an annual cost of around $15,000 (for lower and higher education). This totals $1.1 trillion per year for the country, fully 5 percent of our nation's economy. While we have brought higher-end technology and sophisticated team environments into many classrooms, the material that is covered has only slightly evolved. In the 2016-17 academic year, 1.96 million four-year college degrees were awarded in the United States. At least 95 percent of the content studied for those degrees was the same content as those awarded twenty years prior. If our job market is changing at a pace exponentially higher than two decades ago, we must adapt our education system to do the same.

---

[12] "More Than 76 Million Students Enrolled in U.S. Schools, Census Bureau Reports," U. S. Census Bureau, last updated December 11, 2018, https://www.census.gov/newsroom/press-releases/2018/school-enrollment.html.

With so many children in so many different areas of study, how can we determine what to teach and where? We must consider the growing areas of our economy, evolution of non-automated jobs, critical required skills, geographic trends, efficient learning environments, required investment, and more. This is a complex system problem that will require much data, organization, analysis, and strategy. But first, we must decide to do it.

In 2020, the coronavirus pandemic and the government lockdowns hit the world's population harder than anything in modern history. It will probably take us years to recover. However, there is a bright side. It showed us some of our society's biggest weaknesses and gave us a glimpse of how to address them. If we learn enough from coronavirus to address even bigger challenges like the ones I mentioned above, in the end, COVID-19 could have a silver lining.

In order to capitalize on these lessons, we will have to make some changes. And changes are not easy, especially when these changes move us away from the ways we have done things for decades. Shifting from a more qualitative decision strategy to quantitative decision processes and decision precision could very well be the next phase in human advancement. Whether you are one of those identifying the objectives, gathering data, designing data organization, creating algorithms, or building the strategies to carry you into the future, or if you are someone who is establishing a culture of quantitative decision making, you have a critical role to play.

Following the examples of organizations like the Oakland A's, GiveWell, and SmartAC, we can achieve things we have only dreamed of by using data and experts in ways we never have before. The world's biggest problems are waiting. When faced with your next Crucial Decision, you may have to make a gut call, but try asking your team, your supervisor, or yourself, "Can we *Reliaball* this?"

Figures and Tables

Bibliography

Campbell, John D., and Reyes-Picknell, James V., UPTIME: STRATEGIES FOR EXCELLENCE IN MAINTENANCE MANAGEMENT, Second Edition. New York: Productivity Press, 2006.

Corbett, Steve, and Fikkert, Brian. WHEN HELPING HURTS: HOW TO ALLEVIATE POVERTY WITHOUT HURTING THE POOR... AND YOURSELF. Chicago: Moody Publishers, 2014.

Fung, Dr. Jason. THE OBESITY CODE: UNLOCKING THE SECRETS OF WEIGHT LOSS. Vancouver: Greystone Books, 2016.

Gulati, Ramesh. MAINTENANCE AND RELIABILITY BEST PRACTICES, Second Edition. New York: Industrial Press, 2013.

Hubbard, Douglas W. THE FAILURE OF RISK MANAGEMENT: WHY IT'S BROKEN AND HOW TO FIX IT. Hoboken: John Wiley and Sons, 2009.

Kahneman, Daniel. THINKING, FAST AND SLOW. New York: Farrar, Straus, and Giroux, 2011.

Newport, Cal. DEEP WORK. New York: Grand Central Publishing, 2016.

Reiter, Ben. ASTROBALL. New York: Three Rivers Press, 2018.